"*The Human and the Machine in Literature and Culture* highlights relationships between human agency and automation in literary imaginations. Its investigation of poetics and poetic production offers fresh insight into the value of the unruly and alive humanity that exists beyond the battery of the machines propelling us toward futurity."

—**Saba Syed Razvi,** Associate Professor English and Creative Writing, University of Houston-Victoria, USA

The Human and the Machine in Literature and Culture

Automation is everywhere: in the supermarket, in home appliances, and on our commutes. While we worry about what automation means for human autonomy now, human societies have long wondered about their replacement by machines. *The Human and the Machine in Literature and Culture* explores the pervasive – and long-standing – influence of automation on humanity by dismantling the prevalent future-oriented perspective of many automation debates. This collection examines how literature has conceptualized automation over centuries, from utopian visions of a world liberated from work and domestic labour to dystopian futures in which humans are surplus to requirements. We set out social and industrial developments which feed into discourses of automation and its mediation in literary cultures. By bringing together theoretical approaches to real-world automation with readings of its literary interpretations, this volume demonstrates literature's role as a space for hypothesizing alternate realities, making clear literature's propensity to inform our attitudes to real-world phenomena.

Kate Foster is a Lecturer in French Studies at the University of Reading, UK. Her research focuses on intersections of human bodies and technology in late nineteenth and early twentieth-century cultures. She is working on a monograph on fictional androids and cyborgs, and developing a new project on technology, disease and cultural history.

Molly Crozier is an early career researcher in French and Comparative Literature. Her research focuses on embodiment, gender and disability in twentieth-century theatre. She is working on a monograph on disability in Samuel Beckett's drama. She holds an honorary fellowship at the Institute of Irish Studies, University of Liverpool.

Perspectives on the Non-Human in Literature and Culture

Series Editor: Karen Raber, *University of Mississippi, USA*

Literary and cultural criticism has ventured into a brave new world in recent decades: posthumanism, ecocriticism, critical animal studies, the new materialisms, the new vitalism, and other related approaches have transformed the critical environment, reinvigorating our encounters with familiar texts, and inviting us to take note of new or neglected ones. A vast array of non-human creatures, things, and forces are now emerging as important agents in their own right. Inspired by human concern for an ailing planet, ecocriticism has grappled with the question of how important works of art can be to the preservation of something we have traditionally called "nature." Yet literature's capacity to take us on unexpected journeys through the networks of affiliation and affinity we share with the earth on which we dwell—and without which we die—and to confront us with the drama of our common struggle to survive and thrive has not diminished in the face of what Lyn White Jr. called "our ecological crisis." From animals to androids, non-human creatures and objects populate critical analyses in increasingly complex ways, complicating our conception of the cosmos by dethroning the individual subject and dismantling the comfortable categories through which we have interpreted our existence. Until now, however, the elements that compose this wave of scholarship on non-human entities have had limited places to gather to be nurtured as a collective project. "Perspectives on the Non-Human in Literature and Culture" provides that local habitation. In this series, readers will find creatures of all descriptions, as well as every other form of biological life; they will also meet the non-biological, the microscopic, the ethereal, the intangible. It is our goal for the series to provide an encounter zone where all forms of human engagement with the non-human in all periods and national literatures can be explored, and where the discoveries that result can speak to one another, as well as to scholars and students.

Natural and Artificial Bodies in Early Modern England
Literature, Natural Philosophy, Objects
Alvin Snider

The Human and the Machine in Literature and Culture
Cultures of Automation
Edited by Kate Foster and Molly Crozier

Contemporary Visual Poetry
Women Writing the Posthuman
Fiona Becket

For more information about this series, please visit: www.routledge.com/Perspectives-on-the-Non-Human-in-Literature-and-Culture/book-series/PNHLC

The Human and the Machine in Literature and Culture

Cultures of Automation

Edited by Kate Foster and Molly Crozier

Routledge
Taylor & Francis Group
NEW YORK AND LONDON

Designed cover image: © Getty Images | DKosig

First published 2026
by Routledge
605 Third Avenue, New York, NY 10158

and by Routledge
4 Park Square, Milton Park, Abingdon, Oxon, OX14 4RN

Routledge is an imprint of the Taylor & Francis Group, an informa business

ISBN: 9781032895871 (hbk)
ISBN: 9781032900452 (pbk)
ISBN: 9781003545910 (ebk)

DOI: 10.4324/9781003545910

Typeset in Sabon
by Deanta Global Publishing Services, Chennai, India

Contents

Contributors

Madeleine Chalmers is a Lecturer in French Studies at the University of Leicester, UK. Her work revives modern French avant-garde writing to engage critically with twenty-first-century questions in the fields of science, technology and epistemology. She is the author of *French Technological Thought and the Nonhuman Turn* (Edinburgh University Press, 2024).

Laura Alice Chapot is a postdoctoral researcher in Comparative Literature and Digital Humanities at Dartmouth College, Hanover, NH. She is working on her first monograph: *Modeling Decadence: Literature, Computation, and Modern Anxiety*. Her research has appeared in *Volupté: Interdisciplinary Journal of Decadence Studies* and the *Canadian Review of Comparative Literature*.

Bruno Ministro is a Junior Research Fellow at the Institute for Comparative Literature, University of Porto, Portugal. His research draws on comparative media studies and digital humanities to explore experimental poetry and electronic literature. He is interested in understanding how literary forms engage with material and social dimensions of media.

Léon Pradeau is a scholar of French and Francophone poetics. His peer-reviewed articles include 'Cécile Mainardi's *Sorties*: Poetry as Desire and Performance' (*Dalhousie French Studies*, 2024) and 'Une époque invalidée? Corps et écriture chez Liliane Giraudon' (*Études Francophones*, 2024). His book, *This Is It,* was published by Antiphony in 2025.

Rebecca Reilly is a PhD candidate in Italian at the University of Cambridge, UK. Her PhD analyses writings by thirteenth- and fourteenth-century Italian authors, examining, through the lens of queer theories of temporality, how these works articulate non-linear constructions of time intertwined with non-reproductive modes of sexuality and gender.

Ben Roberts is an Associate Professor in Digital Humanities at the University of Sussex and Co-Director of the Sussex Digital Humanities Lab. He has published widely on the philosophy of technology, particularly the work of Bernard Stiegler, media archaeology and automation.

David Spieser-Landes is an Associate Professor of French and Cultural Studies at the University of North Carolina Wilmington, NC. His primary research areas include postcolonial studies, nation studies and minority studies. Prior to joining the faculty at UNCW, he taught at Penn State University, Lycoming College and the University of Pittsburgh.

Emanuele Stefanori is an Adjunct Lecturer of Italian at Carnegie Mellon University (USA). He earned a PhD in Romance Studies from the University of North Carolina at Chapel Hill. His research explores the role of technology and the agency of non-human actors in Italian crime fiction.

Vanessa Weller is a PhD candidate in French and Francophone Studies at Michigan State University, East Lansing, MI. Vanessa is the Editor-in-Chief of *TROPOS*. Her work has appeared in *Magistra: A Journal of Women's Spirituality in History*. Her dissertation discusses the intersections of class, race and gender in the twentieth-century Parisian avant-garde.

Acknowledgement

This project has been through many iterations: from its first days as the beginnings of an ever-evolving postdoctoral project, to an interdisciplinary conference, to this edited collection. Going back further, the germ of this project has its roots in an early recognition that our research spoke to each other's in multiple ways, which itself led to a joint research seminar on the prosthesis. Human bodies and their relationships with technology have long been of interest for us both.

There are many people and organisations to whom our thanks are due. Firstly, the speakers from the interdisciplinary conference 'Automation and Automatism/s', held in January 2023 at King's College London: Emanuele Stefanori, Anita Barmettler, Howard Ryland, Thomas Crew, Ben Roberts, Barbara Cohen Bastos, Elizabeth Beckman, Sylvie Durmelat, Jess Majekodunmi and Eleonora Lima. Your contributions made for a fantastic conference and helped to shape our thinking and planning for the present volume. Friends and colleagues helped with the practical elements of conference organisation: thank you, Adina Stroia, Alexandra Pugh and Jonathan Hann, for giving up your time to support us. We are grateful to King's College London for hosting the conference and to their AV team for facilitating our plan to hold a hybrid event on a Saturday. We owe our thanks to the organisations who kindly funded the conference: the Society for French Studies, the Society of Dix-Neuviémistes and the Department of Languages, Literatures and Cultures at King's College London. Thank you to colleagues whose thoughtful input and comments have influenced the project at different points, particularly Susan Harrow, Johanna Malt, Emma Bielecki and Luca Viganò. Finally, our warmest thanks to the French Department at King's College London, which provided an open and collaborative space for its doctoral researchers – literally open in the form of the glass-walled doctoral office we shared (the fishbowl) – and in which students form a key part of the department's research community. For so many reasons, this book would not have happened without you – *merci infiniment*.

Kate Foster and
Molly Crozier

London & Newcastle, March 2025

Introduction

This Time It's (Probably Not) Different

Kate Foster

'Industrial robot crushes man to death', proclaimed *The Guardian* in November 2023. A robot at a distribution centre in South Korea reportedly mistook the worker for a box of produce it was supposed to be stacking, and the man was fatally crushed against the conveyor belt (*Guardian* 'Industrial Robot'). While this seems to be a case of malfunction or faulty programming rather than some malevolent intent on the part of the robot, it nevertheless feeds into a long-standing narrative of apprehension about automated machinery. Sometimes, as in the South Korean case, this comes from fear of physical harm, evidenced by another *Guardian* article, which reported that Tesla's autopilot software had been involved in more than a dozen serious incidents. According to the US National Highway Traffic Safety Administration, Tesla's Autopilot (a partially automated driving system) did not include sufficient safeguards to prevent drivers' overreliance on the software and was implicated in 13 fatal crashes and several more involving serious injury (*Guardian* 'Tesla's Autopilot'). Resulting more from failures of or gaps in programming than from automated technologies possessing 'a will that is in conflict with our own', as Stephen Hawking predicted for the future of AI during a speech in 2016 ('The Best or Worst Thing to Happen to Humanity'), these events where humans are harmed or worse by machines nevertheless seem to evoke the creations of sci-fi imaginings. Are we now witnessing the more realistic – because more mundane – scenario of how technology will be the ultimate undoing of the human species in such errors and unanticipated consequences? Or are these examples merely the prelude to a dystopia in which the machines wilfully attack and destroy us, *Terminator*-style?

More prevalent than fear of physical harm in cultural discourses around automation is the concern that automation will lead to job losses. Precisely this fear was behind a threatened strike by US dock workers, who reached an agreement with employers in early 2025 to safeguard jobs whenever a semi-autonomous crane is introduced. As a result, one additional worker will have to be employed for each new crane (Berger 2025). Sometimes, the

DOI: 10.4324/9781003545910-1

reasons for anti-automation sentiments are not so directly linked to jobs. As Laura Chapot outlines in these pages, the Luddites – often invoked as the poster children for technophobia – were in fact protesting against machines which undermined quality and disrupted collaboration, rather than against machines across the board. Yet a rise in automation – predicted or actual – does not always arouse negative responses. In May 2024, for example, AI company Wayve secured $1.05 billion of investment for its AI-powered self-driving vehicle project ('Vote of Confidence'), a signal of investors' faith in this kind of technology (albeit still in development in 2025). Elsewhere, food production and distribution are at the forefront of automation in the form of driverless forklifts, tractors and other robots. At a plant outside Los Angeles, up to 725,000 kilogrammes of lemons are squeezed per day, the juice from which is shipped to Chick-fil-A restaurants to produce its famous lemonade. This operation replaces 10,000 human hours worked daily across Chick-fil-A locations, where employees previously squeezed lemons by hand (De Sousa 2025). While this development might seem to threaten jobs, Agnieszka De Sousa's article suggests that large-scale deportations, which may take place under the second Trump administration, will leave the US agriculture and food production sectors with significant worker shortages, which automation could help to alleviate (De Sousa 2025). This contemporary example of food production is emblematic of the historic links between automation and marginalized humans highlighted by Kevin LaGrandeur (2012) and Despina Kakoudaki (2014), whose focus is the connection between automation (in the form of artificial humanoids) and slavery. LaGrandeur suggests that humanoid bodies have often been designed as artificial slaves in order to perform work which human subjects would rather avoid. While LaGrandeur traces this desire for non-human workers as far back as Aristotle, it is possible that emancipation laws in nineteenth-century Europe intensified the wish for artificial servants to carry out the work previously done by slaves. The relationship between US food production and slavery is, of course, well-known. Today, some seek to use automation as a way of justifying anti-immigrant politics: the UK's shadow Home Secretary, Chris Philp, suggested that robots could be used to replace migrant workers to reduce net migration (Courea 2024). The notion of technology replacing certain types of workers is developed by Neda Atanasoski's and Kalinda Vora's (2019) concept of the 'surrogate relation', in which technology is inevitably founded upon unseen and unacknowledged labour, an idea explored in Bruno Ministro's chapter in the present volume. More broadly, race has long been enmeshed with technology, as outlined by Stephen Cave and Kinta Dihal (2020). They convincingly argue that AI (or technology) is often racialized as white due to the normalization of whiteness in the West, in which technological superiority (in the form of transportation, weapons and more) was used to justify colonialism. Furthermore, they suggest that coding AI/technology as white allows the white 'master' to avoid uncomfortable interactions with people of other races. This complex intersecting of race, automotive technology and labour relations is key to David Spieser-Landes' chapter here. While the future iterations of automation cannot be fully known, it seems certain that automation will alter the landscape of employment significantly.

What Is Automation?

But what do we mean by automation? Are we all talking about the same thing? In the first chapter of this collection, Ben Roberts interrogates the history of the word 'automation', which was popularized from the mid-twentieth century onwards. These days, definitions of automation tend to run something like this: the performance by machines of tasks normally or previously performed by humans. Some definitions also include the criteria that this activity take place within a self-governing system, thus distinguishing automation from mechanization. Automation might also be described as the use of technology/software/robotics (delete as applicable) to carry out tasks with less or little human involvement, or making a process or tool work automatically. What such definitions share is an emphasis on concepts such as 'task' or 'work', a sense that automation necessarily (automatically?) involves the replacement of human labour, paid or otherwise, by machines and/or technology. Certainly, examples of automation such as washing machines, warehousing robots and numerically controlled machine tools are labour-saving devices, reducing the amount of human time (and often money) required to perform the same task.

Perhaps that is why automation frequently evokes fear and apprehension, because it tends to be associated with the loss of jobs. Indeed, there are whole books devoted to precisely this subject.[1] Yet there are alternative perspectives on what automation will mean for the world of work. Journalist John Aziz argues that, as we enter the second quarter of the twenty-first century, we are not living under late-stage capitalism, but at the dawn of 'autocapitalism'. In this new age, Aziz claims, there will be significantly fewer people employed as wage labourers as humans will increasingly make money as capitalists, through asset ownership, investment and so on (2025). For Aziz, automation will fundamentally change capitalism, perhaps eventually leading to a world free from work, as envisioned by Aaron Bastani in *Fully Automated Luxury Communism: A Manifesto* (2019). In terms of machine intelligence, Catherine Malabou (2019) argues that machines and technology are now mimicking neural synapses in the human brain, thus showcasing technology's plastic potential. As she explains, the traditional opposition of technology and the human (brain) is not just overly simplistic but profoundly unhelpful for any attempt at resisting automation (114–5). Madeleine Chalmers' chapter in this volume calls for a similar rethinking of the distinction between human and machinic cognition. Automation and automatism, Malabou reminds us, have always had a double meaning, of involuntary movement and of something moving by itself, giving the lie to assumptions that automation necessarily negates agency (Malabou 100–1). For Malabou, automation and technology are not discourses of control (over us); rather, we can use automation and technology to let go of control, an act which is liberating in itself. Elsewhere, predictions about the decline of paid human work, which might be expected under full-scale automation, have other positive aspects. For some thinkers, job losses, while inevitable, could free humans to pursue other activities such as hobbies, artistic creation and more.[2] Such a development would represent a continuation of movements to reduce the length of the working week in the early decades of the twentieth century, as documented by Benjamin Kline Hunnicutt (1988). For Andy Stern and Lee Kravitz, a universal basic income is the sensible answer to the practicalities of a workless future (2016), and back in 1964, a UK Labour Party manifesto emphasized how leisure time would take on greater significance in our lives as a result of automation ('The New Britain'). Such thinking can be conceptualized as 'a flip in production: automation as the ultimate refinement of standardized consumer production paradoxically opens the door to the transcendence of capitalism *and its consumerist values*' (Bassett and Roberts 14, emphasis original).

But what about automation unrelated to the concerns of the working world, as seen in automata produced as curiosities? Jacques de Vaucanson's flautist, for example, delighted and perhaps unsettled Parisian audiences in the 1730s by playing 12 different melodies and, thanks to bellows placed inside, even seemed to breathe (Wood 22). Later in that century, clockmaker Pierre Jacquet-Droz created automata, including the Writer, capable of writing any text up to 40 letters long (not quite as long as the original Twitter character limit of 140) and the Draughtsman, which resembles a child and can draw four different pictures (Jacquet-Droz.com). These machines were ends in themselves, created to entertain audiences and to draw attention to the technical skill of the makers behind them. They had no practical function and did not carry out tasks previously or typically performed by humans. Such creations have fictional counterparts, including Villiers de l'Isle-Adam's Hadaly (*L'Ève future*, 1886), Filippo Tommaso Marinetti's eponymous *Mafarka* (1909) and Gaston Leroux's Gabriel (*La Poupée sanglante*, 1924), to name but a few. Such examples of automation and automated humanoids defy the definitions discussed above which, in their emphasis on the replacement of human labour, foreground the usefulness and, consequently, the financial profit to be drawn from automation. Such definitions, then, founded on a premise of capitalist logic which insists upon the use value of objects and processes, do not make room for items created for pleasure, entertainment or simply because their makers *could*. These definitions also tend to assume that automation is a twentieth- and twenty-first-century phenomenon and that it is somehow always linked to cybernetics and electronics. Indeed, in the 1950s, the word automation had not yet made its appearance in the dictionary (Bassett and Roberts 11), although the OED gives uses of the word going back to the 1870s. But just as machines and technology are not inventions of the Industrial Revolution, neither is automation, even if the word did not exist before that time.

As I have argued elsewhere, an ambiguous example of automation and its relationship to capitalism exists in Émile Zola's *Au Bonheur des Dames* (1883) in the form of the titular department store's chief cashier, Lhomme. One-armed Lhomme counts the store's takings, 'with the speed and precision of an automated machine', using his remaining arm, yet his superhuman abilities have reduced him to a spectacle which customers come to watch. Lhomme's injury is caused by technology (an omnibus), but this is not an example of man being replaced by machine. Lhomme becomes machine-like and objectified as a result of his injury but still derives pleasure from counting cash. His automatization, therefore, is not wholly dehumanizing and neither is it solely reducible to money and usefulness, despite what his occupation might suggest (Foster 85–6). A further limit case in terms of use value is to be found in the 'Robot Artist' Ai-Da. Built in 2019 by Engineered Arts and conceived by Aidan Meller, Ai-Da is a humanoid robot which 'creates' its own paintings, drawings and poetry. As no two works by Ai-Da are the same – a result of the innumerable possible outcomes of different brush strokes, colours and so on – the robot is said to be producing art (Davies 2022). Originality, of course, is not the same as creativity, and many might hesitate to describe Ai-Da's works as art. Ai-Da was created ostensibly as a research project to pose ethical questions about technological change, machine learning and AI algorithms (Davies 2022) and is not intended to replace human labour in any way. As such, Ai-Da seems to be free from the frequent association of automation with human labour. Indeed, Ai-Da answered questions as part of a UK House of Lords inquiry on 'Creative Futures', which led to a bizarre exchange: 'I should have said at the start, the robot [Ai-Da…] is providing evidence, but it is not a witness in its own right'. Pausing briefly as several laughs erupted around the room, the chair added, 'I don't want to offend the robot' ('A Creative Future'). Asking an AI robot to give evidence about itself (through pre-submitted questions) is not directly related to financial concerns, yet part of the inquiry's focus was the impact of AI and other automated technologies on the creative industries and, by implication, on jobs in those areas. (One might conclude that the decision to invite Ai-Da to speak was motivated more by a desire to shake off the upper house's reputation for being out of touch through a session guaranteed to garner media attention, rather than any serious attempt to understand AI's effect on the creative industries.) On the other hand, Ai-Da and the work 'she' produces have proved to be financially viable for someone at least: a painting of Alan Turing by Ai-Da was sold by Sotheby's for more than $1 million in November 2024 (Porterfield 2024). Ai-Da's liminal status as profitable by conventional capitalist logic yet only incidentally so complicates Jean Baudrillard's distinction between automaton and robot. For Baudrillard, 'the automaton is the *analogon* of man' ('l'automate est l'*analogon* de l'homme', emphasis Baudrillard's), whereas the robot is a worker (82–3).[3] This emphasis on work chimes with the etymology of robot, a word famously coined by Karel Čapek in his play *R.U.R.* (1920), in which robot workers revolt against their human overlords. 'Robot' comes from the Czech word 'robota', or forced labour (although Čapek's robots have organic flesh and are thus closer to what we would call cyborgs today). While both Čapek and Baudrillard associate the word robot with work, for Baudrillard, the automaton's purpose is to pose questions about the nature of human existence, and about the difficulty of distinguishing truth from appearance (Baudrillard 82–3), much like the claims made for Ai-Da. While we do not dispute the importance of automation in the world of work, we suggest that the centrality of labour and employment in automation debates is unnecessarily narrow and, furthermore, that such a focus on employment tends to tie concepts of automation to industrialization, mechanization and digitization and, consequently, to a post-1750s society. We contend that by expanding our definition/s of automation, as well as the timeframe in which we understand it to be taking place, we can uncover repeated tropes and perspectives, leading to a more nuanced understanding of a future which seems likely to be increasingly populated by technology.

If Vaucanson's automata and any number of fictional automated machines subvert or deny the emphasis on the value to be obtained through automation, the relative visibility of automation and human workers also troubles many mainstream definitions of automation. High-frequency trading (HFT) – which uses computer algorithms to carry out a large number of transactions in a second or less – has received attention from N. Katherine Hayles, among others, who alleges that this type of automated trading, which is only possible because of the faster processing time of computers compared to human traders, unfairly contributes to market volatility (143, 167–8). In academia, there has been much chatter about students using ChatGPT or other large language models (LLMs) to write some or all of their essays (Adams 2024). These technologies are examples of automation which replace human labour but often clandestinely or, according to some analyses, unfairly. Allegations of unfairness – academic cheating or financial misconduct – rely less on perceived rule-breaking and more on the assumption that in at least some cases, transactions are being carried out or essays are being written by actual human people. In this way, the unfairness rests on what is perceived to be an unlevel playing field, that the computer-driven automation is necessarily superior to the same process when it is carried out by a human.

On the flip side of this is automation which obscures the human labour behind it, where the automation is akin to a smokescreen behind which are hidden any number of human labourers. The Alfred Club, an app-based service which allows subscribers to book 'Alfreds' to complete domestic tasks when the user is away from home, depends on nameless humans to actually carry out the tasks. Yet, as Atanasoski and Vora have observed, the Alfred Club's USP is precisely the automatic, non-contact nature of the digital platform via which the user orders the service, which means that any 'Alfred' is interchangeable with any other, and with whom the subscriber needs to have no interaction whatsoever. As Atanasoski and Vora note, the lack of contact with or personal knowledge of any one Alfred makes it very much easier to pretend that it is not a real person washing one's dirty socks (2019 93–4). Similarly obfuscating of human workers is the Amazon Mechanical Turk Marketplace (AMT), a kind of crowdsourcing employment platform where users post tasks to be completed by humans for a given price (discussed in detail in Ministro's Chapter 8). Atanasoski and Vora again invoke 'the surrogate effect', this time linking it to AMT, where, like the Alfred Club, the human workers are anonymized and interchangeable to the point of invisibility. What looks like artificial intelligence is no such thing: it is '*artificial* artificial intelligence' (2019 99). The huge numbers of gig workers able to complete tasks at any time of day or night, and who are willing to do so for very small amounts of money, 'replicat[e] the imagined speed of digital processors' in circumstances when developing an algorithm would be too time-consuming (2019 100). For Atanasoski and Vora, these invisibilizing automated technologies are a preliminary step en route to full automation in which humans will be increasingly sidelined.

How New Is It?

This all sounds like a very twenty-first-century problem. But is it? While automation seems to be playing an increasingly significant role in our lives, from supermarket self-checkouts to chatbots, from factory robots to driverless cars, our relationship with automation is not a new one. The twentieth century witnessed numerous developments in the world of automation which tend not to arouse significant debate (at least not today), including refrigerators (1913), central heating thermostats, and Henry Ford's moving assembly line (1913). The nineteenth century birthed automation innovations such as washing machines (1851), sewing machines (1829), trains (1802) and the Jacquard loom (1801), while Charles Babbage's difference engine came into being in the 1820s (for more on this see Laura Chapot's chapter in these pages). And before that, examples of automation included engines which used steam for practical reasons (eighteenth century), Johann Gutenberg's movable-type printing press (1450s), windmills with automatically turning sails (developed in the Middle Ages), Ismail al-Jazari's automata, which included fountains and automated servants (twelfth century), novelty steam engines (first century CE) and Hero's trumpet automaton (a first-century automated doorbell).

Clearly, some of these developments were curiosities produced for the pleasure of making or owning such a device, while others replaced jobs previously carried out by human workers. Yet they did not bring about mass unemployment or the collapse of capitalist society: the invention of the sewing machine – particularly the commercially successful model produced by Singer in the 1850s – combined with cheaper fabrics due to mass production, lowered the price of clothing and pushed forward the development of ready-to-wear fashion and the department store where it was sold (Harvey 218; Higonnet 353). This in turn led to a raft of new jobs within the stores and in the ecosystem which supported and enabled them, including construction, delivery drivers, the producers of the innumerable commodities which they sold, not to mention the workers required to build, maintain and repair the sewing machines themselves. While many worry today about the impact of automation, that tens of thousands of jobs will be lost as we struggle to adapt to what feels like unprecedented technological change, this is nothing new. As Cynthia Estlund observes, automation demonstrates a historic tendency to destroy some jobs while creating others, new jobs which are typically better remunerated and less physically demanding (12). This, of course, is not to suggest that the loss of any job/s, even if others are created elsewhere, is any less devastating on a personal and community level. On the other hand, we might reasonably hope that increased automation will bring greater safety and comfort, minimizing accidents and repetitive tasks in our daily lives. This has already been borne out in a number of ways, as home appliances such as dishwashers have reduced the monotony of domestic tasks (to some extent at least), autopilot features in aeroplanes enhance navigation during adverse weather conditions and, as Estlund notes, care robots can alleviate some of the more onerous physical tasks required in care work (7). While discourses of automation tend to focus on the present and the future, *The Human and the Machine in Literature and Culture* dismantles this future-oriented gaze and places automation debates in a historical and cultural context. This volume proposes that literature and literary criticism are privileged sites where anxieties and hopes surrounding automation have long been played out and problematized.

Cultures of Automation

At the heart of automation is the interface between human and machine. As we have seen, automation today is everywhere: in the supermarket, in home appliances and on our commutes. While we worry about what automation means for human autonomy now, human societies have long wondered about their replacement by machines. *The Human and the Machine in Literature and Culture* explores the pervasive – and long-standing – influence of automation on humanity by dismantling the prevalent future-focused perspective of many automation discourses. We do not take a chronological approach to the material in this volume, as we do not wish to suggest that human society is engaged in a unidirectional march towards an automated future. Instead, we follow Caroline Bassett and Ben Roberts in proposing that automation debates recur through time, that 'automation anxiety', together with 'automation fever' (enthusiasm and utopian predictions about its potential), are 'cyclical phenomen[a]' (Bassett and Roberts 9). These themes feed into our central contention that automation debates are always reflective of other concerns of the time and location in which they take place, which the historical and international survey of our chapters brings to the fore. In the 1960s, for example, automation anxieties on the political left were fed by fears that, in a work-free automated world, the ability to voluntarily withdraw one's labour would entail reduced dignity for the working man (Bassett and Roberts 15), a pertinent concern in a decade characterized by workers' strikes, from France's *mai '68*, the UK miners' strike (1969) and the Dodge Revolutionary Union Movement (formed in 1968) at Chrysler's assembly plant in Detroit. Earlier, in 1907, gas lamplighters in New York City downed tools in protest at the Edison lightbulb, which they feared would put them out of work (*New York Times*, 'Lamplighters Quit'). Before that, in 1858, *The New York Times* fretted that the brevity of telegrams would encourage a lack of depth and even misinformation, that the young would become addicted to this new form of messaging and that it was largely pointless to be able to communicate such short messages so quickly (*New York Times*, 'The Ocean Telegraph'). One might easily replace 'telegram' with 'Twitter' in the preceding sentence, for the fears about misinformation and the telegraph did in fact come to pass. In one example, telegrams falsely reported that Jefferson Davis was in personal command of the Confederate Army (Sternhell 2013). In short, discourses of automation are both culturally and historically contingent.

This collection examines how literature and culture have conceptualized automation over centuries, from automata which seem to reinforce our own mortality, to poetical forms influenced by computation and prostheses, to positive understandings of LLMs, and dystopian fears of a world where humans are surplus to requirements. In many ways, literature provides a space for experimentation with new ideas, free from real-world constraints, and is often futuristic in setting or conception. As subject matter, automation in literary cultures provides a connection to the real world, yet can often remain somewhat fantastical, bridging a gap between the relatability of the familiar and a fictional world which does not quite exist in reality. In this way, automation in literature is reflective of the social preoccupations of the time and culture in which it is produced. It can thus uncover something of those preoccupations and the feelings which automation has evoked at different points. Literary examples of automation are frequently thought experiments which depend for their meaning upon a technology which had not been (fully) realized at the time of writing. By looking back at these examples – perhaps even from a time by which some of these inventions and developments have come to pass in one way or another – we can begin to identify patterns in human responses to automation, allowing for more measured and informed reactions today and tomorrow.

To reverse this proposition, we might ask how automation can help us to read and interpret literature. What results might be gleaned from applying automatic and machinic processes to literary analysis? In several of the chapters in this volume, reading literature through the lens of automation yields new insights into both form and content, uncovering how pervasive automatic modes of thinking can be and how intricate and entwined is the relationship between human and machine.

Automation Debates

The last decade has witnessed a raft of publications on automation in the twenty-first century, largely focusing on its predicted repercussions on employment. Richard and Daniel Susskind argue that it is not only lower-skilled workers who are at risk from automation but that automation is coming for professions such as lawyers, accountants and doctors. This, they suggest, is a result of a wholesale – if gradual – move from a 'print-based industrial society' to a 'technology-based Internet society', in which the high price of the services offered by professionals, together with their tendency to keep knowledge inaccessible to most, are fundamentally undemocratic and therefore ripe for change (Susskind and Susskind 3, 42–6). Richard Susskind develops some of these ideas elsewhere, noting that 'automation anxiety' began in the wake of the Industrial Revolution as people feared machines would supplant jobs: as he observes, automation anxiety and economic expansion were linked from then on (Susskind, *A World Without Work* 14-15). Basing his position on the premise that we are in the midst of a period of 'labour underdemand [sic]', Aaron Benanav argues that this reduced requirement for workers is not (only) the result of automation (9). Cynthia Estlund, a professor of labour law, cautions that 'this time [is] different', that the imminent iteration of automation will, unlike those in the past, result in wide-reaching job losses (Estlund 1). For Estlund, this matters not just in terms of the economic impact (not to mention the personal impact for the unfortunates who lose their jobs and may not be able to reskill in order to be redeployed elsewhere in the economy), but because, with the loss of useful paid work, something more fundamental will be lost to human society and human subjectivity.

While these texts highlight a broader cultural relevance surrounding automation debates today, we are concerned here with humanist and cultural responses, both now and in the past. Recent scholarship in cultural studies, philosophy and digital humanities is turning its attention to new understandings of automation and the technologized human, highlighting our fascination with these ideas. Joel Mokyr et al. (2015) highlight the cultural anxiety engendered by technological advances (loss of jobs, the dehumanizing effect of repetitive work and of the loss of work itself) both today and in the past, asking 'is this time different?', chiming with Bassett's and Roberts' observation about the cyclical nature of automation debates. Mark Andrejevic et al. (2023) examine the relationship between automation and culture, arguing that automated systems are increasingly involved in cultural outputs, including music, art and media. Continuing the cultural theme, Linda M. Austin (2018) argues that photography and dance in the final decades of the nineteenth century were responding to debates about automatism from the field of human psychology, arguments which were contentious precisely because they seemed to undermine notions of human agency. Illustrating the more positive side of automation discourse, Catherine Malabou (2019) argues that agency and freedom are not separate from or external to the machine, instead promoting an emancipatory surrendering of control afforded by technology and automation. Madeleine Chalmers (2020) traces a genealogy from Bernard Stiegler's notion of twenty-first-century society as automatic (and which goes beyond the replacement of humans by machines) back to Surrealist automatic writing. For Chalmers, while some warn that we are 'sleepwalking' into a nightmarish automated future, the power of automatism and automation can be harnessed as a source of creativity and subjectivity with which to navigate our automated society. Similarly upbeat are Kristiina Korjonen-Kuusipuro and Adam Wojchiechowski (2023), who respond to recent developments in AI by emphasizing the need for hope, in combination with ethical guidelines, as we move into a future increasingly dominated by automated technologies.

As machines automate progressively more activities once carried out by people, the human, one might think, risks becoming an obsolete category. Does the future of humankind lie in refashioning ourselves as ever-more machine-like, ever-more automated and unthinking? If an essential distinction between human and machine is conscious thought, when human behaviour is no longer motivated by consciousness, what results is automatism. How might automatism trouble notions of agency in relation to the racial and class dynamics of labour? We seek to answer these questions in the chapters which follow. By teasing out the complexities of automation and its history, this volume will nuance the often-polarizing reactions to automation, which characterize it as either the beginning of the end of human subjectivity or the dawn of a new age in which humanity will be liberated from the monotony of work.

The volume's themes include the intersections of humans and machines (human 'machines', boundaries between living and non-living, non-human actors and humanoid robots); agency in relation to automatic behaviours and practices (artistic and creative processes, empathy and the erasure of human labour); automatism, myth and the uncanny. In Chapter 1, Ben Roberts questions the novelty of automation, as both a word and a concept, in the second half of the twentieth century, and sets out the centrality of anxiety to cultural discourses of automation through his case study of the BBC Horizon documentary 'Now The Chips Are Down' (1977). Emanuele Stefanori (Chapter 2) examines automation and automatism through Italian detective fiction of the 1930s, arguing that conventions of genre automatically generate plot twists, that mechanical elements are humanized and human actors rendered automaton-like, and that non-human technological objects are afforded an agency on a par with that of human characters. In Chapter 3, Rebecca Reilly explores automation as a conceptual tool in medieval Florentine poetry (when real automata remained rare), which both sets out and troubles binaries, particularly, in this case, that between life and death. Vanessa Weller (Chapter 4) examines the intersections of automation and folklore through nineteenth-century German and French fiction, asking how the psychological states of protagonist and/or author are mediated through and affected by technology. Laura Chapot (Chapter 5) interrogates how decadent poetics (in German and Swedish texts) are informed and influenced by computational understandings of the period, arguing for algorithmic and modular forms which produce disorientation. In Chapter 6, Léon Pradeau develops his notion of 'prosthetic verse', which, he argues, describes the work of contemporary French poets who attend to biotechnologies and whose practice frequently deconstructs itself, thus bridging a gap between the reality and the metaphor of prosthesis. David Spieser-Landes (Chapter 7) examines postcolonial French 'beur' writing in which second-generation immigrant authors are able to resist the automatic cultural reproduction which tends to centre narrow definitions of 'Frenchness'. In Chapter 8, Bruno Ministro puts Amazon's Mechanical Turk system under the microscope, through close analyses of literary works which unsettle its narrative of pure automation by exposing the human labour underpinning it. Finally, in Chapter 9, Madeleine Chalmers calls for new ways of thinking LLMs, specifically through the work of Claude Lévi-Strauss and his concepts of *bricolage* (assemblage) and wild thought.

Through this wide-ranging and transnational survey, we suggest that automation is not solely a late-capitalist issue and that thought experiments in literary cultures demonstrate how our fears and hopes around automation may have more basis in imagination than in reality. *The Human and the Machine in Literature and Culture* asks how these understandings of and reactions to automation can help us understand our technological present and our increasingly technologized future.

Notes

1 See, for example, Richard Susskind and Daniel Susskind (2015), Benanav (2020), Johannessen and Sætersdal (2020), Estlund (2021) and Susskind (2021).
2 See, for example, Didier de Chousy's 1883 novel *Ignis*, in which robot slaves carry out the work previously done by humans, who themselves live a life of leisure in a utopian city.
3 Author's own translations.

References

Adams, Richard. "More than Half of UK Undergraduates Say They Use AI to Help with Essays." *The Guardian*, 1 Feb. 2024, https://www.theguardian.com/technology/2024/feb/01/more-than-half-uk-undergraduates-ai-essays-artificial-intelligence/. Accessed 20 Jan. 2025.

Andrejevic, Marc, Robbie Fordyce, Luzhou Li, and Verity Trott. "Automated Cultures: Introduction." *Cultural Studies*, vol. 37, no. 1, 2023, pp. 1–19.

Atanasoski, Neda, and Kalindi Vora. *Surrogate Humanity: Race, Robots, and the Politics of Technological Futures*. Duke UP, 2019.

Austin, Linda M. *Automatism and Creative Acts in the Age of New Psychology*. Cambridge UP, 2018.

Aziz, John. "This Is Not Late-Stage Capitalism." *Quillette*, 8 Jan. 2025, https://quillette.com/2025/01/08/this-is-not-late-stage-capitalism-automation-ai-robotics/. Accessed 16 Jan. 2025

Bassett, Caroline, and Ben Roberts. "Automation Now and Then: Automation Fevers, Anxieties and Utopias." *New Formations*, vol. 98, 2019, pp. 9–28.

Bastani, Aaron. *Fully Automated Luxury Communism: A Manifesto*. Verso Books, 2019.

Baudrillard, Jean. *L'échange symbolique et la mort*. Gallimard, 1976.

Benanav, Aaron. *Automation and the Future of Work*. Verso Books, 2020.

Berger, Paul. 'With Port Strike Averted, Dockworkers Draw New Curbs on Automation." *The Wall Street Journal*, 9 Jan. 2025, www.wsj.com/articles/with-port-strike-averted-dockworkers-draw-new-curbs-on-automation-97938142/. Accessed 14 Jan. 2025.

Cave, Stephen, and Kinta Dihal. "The Whiteness of AI." *Philosophy and Technology*, vol. 33, 2020, pp. 685–703.

Chalmers, Madeleine. "Living As We Dream: Automatism and Automation from Surrealism to Stiegler." *Nottingham French Studies*, vol. 59, no. 3, 2020, pp. 368–83.

Courea, Eleni. "Use Robots Instead of Hiring Low-Paid Migrants, Says Shadow Home Secretary." *The Guardian*, 28 Nov. 2024, https://www.theguardian.com/uk-news/2024/nov/28/use-robots-instead-of-hiring-low-paid-migrants-says-shadow-home-secretary/. Accessed 28 Mar. 2025.

"A Creative Future." Oral Evidence Presented at the House of Lords Communication Committee, 11 Oct. 2022, https://committees.parliament.uk/event/14659/formal-meeting-oral-evidence-session/. Accessed 16 Jan. 2025.

Davies, Caroline. " 'Mind-blowing': Ai-Da Becomes First Robot to Paint Like an Artist." *The Guardian*, 4 Apr. 2022, https://www.theguardian.com/technology/2022/apr/04/mind-blowing-ai-da-becomes-first-robot-to-paint-like-an-artist/. Accessed 16 Jan. 2025.

De Sousa, Agnieszka. "Driverless Farm Machines and Robots: Food's Tech Push Gathers Pace." *Bloomberg UK*, 10 Jan. 2025, https://www.bloomberg.com/news/newsletters/2025-01-10/global-food-roundup-robots-and-farm-machines-highlight-automation-push/. Accessed 10 Jan. 2025.

Estlund, Cynthia. *Automation Anxiety: Why and How to Save Work*. Oxford UP, 2021.

Foster, Kate. "Lèche-vitrines: Human Identity and the Mannequin in *Au Bonheur des Dames*." *Dix-Neuf*, vol. 26, no. 2, 2022, pp. 74–90, https://doi.org/10.1080/14787318.2022.2085021.

Guardian. "Tesla's Autopilot Feature Was Involved in 13 Fatal Crashes, US Regulator Says." 26 Apr. 2024, https://www.theguardian.com/technology/2024/apr/26/tesla-autopilot-fatal-crash/. Accessed 16 Jan. 2025.

Harvey, David. *Paris: Capital of Modernity*. Routledge, 2003.

Hawking, Stephen. Speech at the Leverhulme Centre for the Future of Intelligence, cited in "'The Best or Worst Thing to Happen to Humanity' – Stephen Hawking Launches Centre for the Future of Intelligence." *University of Cambridge Research News*, 19 Oct. 2016, https://www.cam.ac.uk/research/news/the-best-or-worst-thing-to-happen-to-humanity-stephen-hawking-launches-centre-for-the-future-of/. Accessed 15 Jan. 2025.

Hayles, N. Katherine. *Unthought: The Power of the Cognitive Nonconscious*. U Chicago P, 2017.

Higonnet, Patrice. *Paris: Capital of the World*. Belknap Press of Harvard UP, 2001.

Hunnicutt, Benjamin Kline. *Work Without End: Abandoning Shorter Hours for the Right to Work*. Templer UP, 1988.

"Industrial Robot Crushes Man to Death in South Korean Distribution Centre." *The Guardian*, 8 Nov. 2023, https://www.theguardian.com/technology/2023/nov/08/south-korean-man-killed-by-industrial-robot-in-distribution-centre/. Accessed 10 Jan. 2025.

Jacquet-Droz.com. "The Draughtsman." https://galeriedesmerveilles.jacquet-droz.com/en.automata. Accessed 27 Mar. 2025.

Jacquet-Droz.com. "The Writer." https://galeriedesmerveilles.jacquet-droz.com/en.automata. Accessed 27 Mar. 2025.

Johannessen, Jon-Arild, and Helene Sætersdal. *Automation, Innovation and Work: The Impact of Technological, Economic, and Social Singularity*. Routledge, 2020.

Kakoudaki, Despina. *Anatomy of a Robot: Literature, Cinema, and the Cultural Work of Artificial People*. Rutgers UP, 2014.

Korjonen-Kuusipuro, Kristiina, and Adam Wojchiechowski. "Hopes for a Better (Techno)Future." *Human Technology*, vol. 19, no. 1, 2023, pp. 1–4.

LaGrandeur, Kevin. *Androids and Intelligent Networks in Early Modern Literature and Culture: Artificial Slaves*. Routledge, 2012.

"Lamplighters Quit; City Dark in Spots." *The New York Times*, 25 Apr. 1907, https://www.nytimes.com/1907/04/25/archives/lamplighters-quit-city-dark-in-spots-police-reserves-out-in-harlem.html/. Accessed 22 Jan. 2025.

Malabou, Catherine. *Morphing Intelligence: From IQ Measurement to Artificial Brains*. Columbia UP, 2019.

Mokyr, Joel, Chris Vickers, and Nicolas L. Ziebarth. "The History of Technological Anxiety and the Future of Economic Growth: Is This Time Different?" *Journal of Economic Perspectives*, vol. 29, no. 3, 2015, pp. 31–50.

Porterfield, Carlie. "Painting by the AI Robot Ai-Da Sells for More than $1m at Sotheby's." *The Art Newspaper*, 8 Nov. 2024, https://www.theartnewspaper.com/2024/11/08/ai-da-robot-painting-sells-one-million-sothebys/. Accessed 16 Jan. 2025.

Stern, Andy, and Lee Kravitz. *Raising the Floor: How a Universal Basic Income Can Renew Our Economy and Rebuild the American Dream*. Public Affairs, 2016.

Sternhell, Yael A. "Lies, Damned Lies and the Telegraph." *New York Times*, 9 Jul. 2013, https://archive.nytimes.com/opinionator.blogs.nytimes.com/2013/07/09/lies-damned-lies-and-the-telegraph/. Accessed 28 Mar. 2025.

Susskind, Daniel. *A World Without Work: Technology, Automation and How We Should Respond*. Penguin, 2021 [2020].

Susskind, Richard, and Daniel Susskind. *The Future of the Professions: How Technology will Transform the Work of Human Experts*. Oxford UP, 2015.

"The New Britain." 1964 Labour Party Election Manifesto, http://labour-party.org.uk/manifestos/1964/1964-labour-manifesto.shtml/. Accessed 22 Jan. 2025.

"The Ocean Telegraph – Relative Benefits and Evils." *The New York Times*, 19 Aug. 1858, https://www.nytimes.com/1858/08/19/archives/the-ocean-telegraph-laying-the-cable.html/. Accessed 22 Jan. 2025.

"Vote of Confidence in UK Economy as British AI Company Wayve secures over $1 billion to Develop AI for Self-driving Vehicles." Press release by the Prime Minister's Office, 7 May 2024, https://www.gov.uk/government/news/vote-of-confidence-in-uk-economy-as-british-ai-company-wayve-secures-over-1-billion-to-develop-ai-for-self-driving-vehicles/. Accessed 14 Jan. 2024.

Wood, Gaby. *Edison's Eve: A Magical History of the Quest for Mechanical Life*. Anchor Books, 2003.

1 'What We Need Is More Automation'

Automation Debates in the Postwar Period

Ben Roberts

The history of the word automation is not without significance for the history of automation itself. In part, this historical interest stems from the word's contemporary usage, where two somewhat conflicting but not opposed trends are discernible. On the one hand, the term is used generically to describe the substitution of machines for human activity (usually, labour). On this reading, everything from contemporary generative AI, self-driving cars and robots, going backwards through time to the postwar emergence of the digital computer and from there to the textile mills and steam engines of the industrial revolution, even as far as windmills and other simple early machines, can be described as 'automation'. Alongside this very sweeping version of what automation means, there are more qualified accounts, such as that provided by Aaron Benanav in his recent *Automation and the Future of Work,* which tie the emergence of automation to the nineteenth century and the idea of machines that can fully, rather than partially, substitute for human labour (5–6).

There is nothing exactly wrong with the generic, sweeping use of the word automation. After all, one might say the fantail windmill, its sails automatically turned into the wind by the effect of side winds on a smaller laterally mounted 'fantail', is as good an example of feedback-based, cybernetic automation as one could wish for. Equally, there are probably good reasons to think, as Benanav does, that a more complete form of machinic replacement of labour does begin to be systematically thought about in the nineteenth century (Benanav cites Babbage and Marx as evidence).[1] However, neither the generic, ahistorical use of the term automation, nor Benanav's more historically influenced account of what he calls 'automation discourse' really reckon with why the word itself does not come into use until the mid-twentieth century. Why was it needed then if automation discourse was already well under way without needing a new word? Why weren't existing words, such as mechanization, enough?

On the other hand, there is also an influential contemporary interpretation of the word automation that ties it narrowly to the postwar era

DOI: 10.4324/9781003545910-2

and what Ronald R. Kline has eloquently described as the 'cybernetics moment'. On this account, the concept of automation is fundamentally shaped by the postwar developments of cybernetics, information theory and the emergence from these of what we think of as the 'information age' (Kline 70–3).

In this chapter, I will argue that both the generic use of the word automation and its narrow tying to a postwar 'cybernetics moment' fail to do justice to the rich history of the emergence of the term, its application to a variety of different economic and industrial postwar developments, as well as its wider relationship with automation culture, debates and imaginaries.

Automation: A New Word?

'What we need is more automation', declared Delmar S. Harder in the summer of 1948. With these words, the then Ford's Vice President of Manufacturing is generally credited with coining the word automation in its current meaning. As *Business Week* observed in October 1955, Harder originally used the term in a rather narrow sense: the need to speed up the process of moving parts from one machine to another ('Special Report: Automation' 78). However its meaning was quickly expanded by others to encompass a new type of machine control, a transformation of factory labour and a new relationship with machines. As a headline in the *New York Times* of the same year exclaimed, 'Automation Aims At New Freedom: Devices That Run Factories Promise to Release Men for Richer Living' (Freeman). Automation in this broader sense no longer meant simply an extension of mechanization, as Harder perhaps intended, but also, according to some, a qualitative shift in the relationship between human and machine. By the mid-1950s, a vigorous debate about this extension of machine control was already under way. Was automation a continuation of a trend in place since at least the beginning of the industrial age, or something entirely novel? Would it destroy jobs, or create them? Would greater machine control dominate humans or liberate them? These essential elements, already in place 70 years ago, still shape debates about automation in our own time, given an apparent new urgency by artificial intelligence and big data. In American public culture, there was immense interest in this new word 'automation'.

But was it, in fact, a new word? Harder's apparent neologism received official recognition in *Webster's Third New International Dictionary* in 1961 and the OED second edition in 1989. However, from the outset, scholars of language debated whether the term was really new. Margaret Marshall pursued this point in a May 1957 article for *American Speech*. She argued that the word had been used in a 1669 philosophical pamphlet by Simon Patrick. Patrick's 'automation of the world' appeared to take the model of the clock to describe the universe as an 'all-inclusive mechanism' set in motion by God (Marshall 150). While later lexicographers argued that automation was here a misprint for automaton (Marshall and Burchfield), by the time of the Oxford English Dictionary third edition, linguists had rediscovered other existing usage, dating from the late nineteenth century, as a synonym for automatism: action that is automatic, mechanical or beyond voluntary control. Automation, it turns out, was never completely a new word. Somehow Harder's shiny new coinage was always already marked by the patina of existing usage.

The rediscovered historical meanings of the word automation also continue to haunt its modern techno-economic usage. Automation as automatism captures the common concern that automation may lead to machines and processes beyond human control. More questionable lexically, but useful in a tenuous philosophical sense, is Patrick's 'automation of the world'. This divine automation, depicting the universe as a mechanism set in motion by God and working towards a higher purpose, hints at the relationship between automation and what we might now think of as rational progress. This is more than the happy accident a misprint might suggest. What is at stake in divine automation is the inevitability or automaticity of progress itself: does automation await, automatically, over the next horizon?

Nonetheless, the linguistic debate at the time and today's dictionaries concur (and even Marshall agreed) that the term, in the sense in which it was being used in the 1940s and 1950s, was novel and that indeed, automation was a 'new word' for the postwar era. There are a few caveats here. Firstly, there is debate about when Delmar Harder and other engineers first began using the word. A letter to the *New York Times* from the editor and writer Burnham Finney, who specialized in writing about the machine tool industry, testified that Harder himself stated that he and other engineers were using the term as early as 1935 and that it was used frequently by Ford and other companies from then on (Finney 168).

Secondly, even if, outside that narrow engineering cadre, automation only reached widespread usage in the postwar era, it is of course closely related linguistically to a variety of existing terms such as automatic and automaton. These terms are all broadly etymologically derived from ancient Greek αὐτόματος meaning 'acting of itself, spontaneous' (Oxford English Dictionary). The word automatic, in particular, was being used to convey the sense of a more perfect or even total machinic substitution for human labour.

For example, in his 1895 novella *The Time Machine*, H.G. Wells's Time Traveller journeys over 800,000 years into the future, where the Thames Valley is now occupied by a small human-like species called the Eloi. The Eloi appear to him to have their material needs met with no sign of labour or visible means of production:

> The several big palaces I had explored were mere living places, great dining-halls and sleeping apartments. I could find no machinery, no appliances of any kind. Yet these people were clothed in pleasant fabrics that must at times need renewal, and their sandals, though undecorated, were fairly complex specimens of metalwork. Somehow such things must be made. And the little people displayed no vestige of a creative tendency. There were no shops, no workshops, no sign of importations among them. They spent all their time in playing gently, in bathing in the river, in making love in a half-playful fashion, in eating fruit and sleeping. I could not see how things were kept going.
>
> (Wells 55)

The Time Traveller initially sees the Eloi lifestyle as evidence of a human Golden Age, a collective 'communist' society supported by what Wells several times describes as 'automatic' means ('automatic civilisation', 'automatic organisation'), in other words, the more perfect replacement of human labour by machines. Only later does he realize that the Morlocks, a second, subterranean human species, are providing the labour to meet the Eloi's needs while simultaneously farming the Eloi for their own consumption.

As Wells's novella might suggest, some elements of what automation might come to mean are already at work in cognate terms like automatic. In his initial, deceptively utopian reading of the Eloi's situation, the Time Traveller is perhaps drawing on something like the nineteenth-century *avant la lettre* automation discourse Benanav alludes to.

So, although the word is new, the concepts being invoked are not necessarily entirely novel. The real question then becomes: what were engineers, economists, thinkers, union leaders and politicians referring to when using the word 'automation' in the 1950s, and how did they think it was meaningfully different from what previously might have been referred to by other similar terms, for example, mechanization?

Automation: A New Concept?

So what is the new word automation being used to describe in the 1950s? One particularly useful source for understanding this can be found in a report entitled 'Automation and Technological Change' issued by the US Congress Joint Economic Committee in 1956, based on hearings in October 1955 investigating the question of automation. The committee heard testimony from senior business leaders at companies such as Ford, General Electric, General Dynamics, the Bell Telephone Company and Studebaker-Parker, as well as union leaders in areas such as communications, electrics, railroads and the chemical industries. It also heard from scientists, academics and management 'experts' on automation such as Walter S. Buckingham, Vannevar Bush and, perhaps most importantly, John Diebold.

It is obviously difficult to summarize the over 300,000 words of testimony from these hearings. One interesting insight into the spread of discussion in the report can be gleaned using the computational technique of topic modelling. Topic modelling is a tool that can be used for 'distant reading' (Moretti) or 'discovery' (Brett) of themes within a large textual corpus by analysing the co-occurrence of terms. Topic modelling software typically results in a series of topics represented by sets of these commonly co-occurring words. In this case, in a pleasingly circular way, it is an automated way of analysing the automation discourse in the 1956 report.

The result of topic modelling the report can be found in Table 1.1. As is usual with topic modelling, I have provided a named interpretation for each topic in the second column, based on the representative words identified by the software and the paragraphs (chunks of text) linked to that topic.

While not wishing to provide strong conclusions merely on the basis of this modelling process, one or two points of discussion emerge from this process of textual discovery. The first is that direct discussion of cybernetics played a very limited role in the committee hearings. It occurs as a representative word in none of the topics identified. In fact, the word cybernetic only appears once in the whole report. Equally, the name of cybernetics' most famous contemporary spokesperson, Norbert Wiener, occurs only twice (once misspelled as Weiner). It's not that the themes of early cybernetics – open- and closed-loop feedback, control and self-regulating machines – were absent from the testimony before the committee, far from it, but rather that they were only one aspect of a wider discussion about automation which was not, on the whole, framed by cybernetics discourse. Instead, the hearings focused on more prosaic questions of what automation was, where it was happening and what, if anything, Congress should be concerned about. As the topics in Table 1.1 suggest, much of the discussion involved applications of automation: in the chemical industry, in telecommunications, in the office and on railroads.

There were also, of course, concerns raised about automation. One important issue here was national competition. Was America falling behind its European competitors, and what could be done about this? There was a particular concern expressed, fraught with significance in the Cold War context, that training in the skills needed for automation might be better in the Soviet Union than in the United States (Joint Economic Committee 9n2) (one might here compare the 1955 committee's concerns with current Trump administration worries about China's advances in AI – *plus ça change, plus c'est la même chose*).

Table 1.1 Topic Modelling of 'Automation and Technological Change', Report of the United States Congress Joint Economic Committee, 1956[2]

Topic Number	*Interpretation*	*Representative Words*
1	Implementing Automation	automation, project, new, information, think, research, complete, group, possible, section
2	Committee Business	committee, hearings, subcommittee, economic, report, joint, study, opportunity, economy, automation
3	Automation, Skills and Unemployment	workers, skills, unemployment, new, job, displaced, seniority, older, problem, automation
4	Capital Costs of Automation	costs, cost, capital, investment, percent, volume, automation, lower, prices, price
5	Concern about Automation	effects, automation, problems, wishful, believe, thinking, concern, ensley, impact, challenging
6	Comparison with Europe	european, labor, productivity, europe, employment, automation, wages, states, united, plants
7	Contemporary Industrial Applications of Automation	industries, industry, automation, petroleum, foreign, automated, applied, companies, competition, entire
8	Automation and Production Process	plant, plants, production, approach, product, safety, brown, problem, flow, process
9	Telecommunications Automation	bell, telephone, dial, dialing, toll, handle, operator, calls, longdistance, years
10	Unions and Automation	union, bargaining, collective, industry, changes, face, important, international, agreement, unions
11	Automatic Transfer Devices	tape, machine, interlock, production, cycle, instructions, machines, automatic, wiring, loading
12	Definition of the Word Automation	word, definition, new, machine, term, automatic, control, extension, automation, technological
13	Automation and Product Design	book, automation, automatically, davis, mr, responsible, remember, design, want, introduced

(*Continued*)

Table 1.1 (*Continued*)

Topic Number	*Interpretation*	*Representative Words*
14	Automation, and US economic freedom	society, social, free, economic, human, world, natural, resources, policies, foresight
15	Chemical Industry Applications	senator, omahoney, flanders, chemical, experience, intelligence, opportunities, proceeds, expand, increases
16	Railroad Applications	yard, railroad, freight, yards, retarder, prior, yardmen, traffic, employees, railroads
17	Changes in Relationship between Labour and the Economy	shifts, new, relations, markets, shift, economy, places, higher, act, gradual
18	Office Automation	clerical, office, electronic, workers, machines, cut, employment, bureau, years, equipment
19	Automation and Mechanization	mechanization, mechanical, day, industry, mechanize, prospects, things, electronics, feature, pointed
20	Automation, Jobs and Unemployment	employment, output, million, jobs, percent, force, gross, year, unemployment, labor
21	Automation Techniques	automatically, material, costreducing, machine, continuous, building, resourcesmechanical, tolerances, mixing, organizing
22	Differences in Automation between Large and Small businesses	small, business, large, big, producer, units, smaller, mass, flexibility, larger
23	Progress towards Automation	progress, belief, going, probably, thing, war, good, produce, 10-years, disclosed
24	Automation and the Workweek	workweek, week,shorter, hours, 32-hour, 40, leisure, 35-hour, prediction, goods

However, a more significant area of worry was in relation to American workers, their working future and industrial relations. The committee heard from both union leaders and economists about the implications for employment. These concerns about employment were not confined to the corridors of power in Washington but formed part of a wider public discourse about automation. Indeed, the year in which the congressional hearings took place, 1955, might well be described as 'peak automation'.

As Friedrich Pollock noted, apart from the hearings themselves, there were public exhibitions on automation in New York, Chicago and Los Angeles that year, and among unionized workers, there was 'considerable anxiety concerning the progress of automation' (Pollock 95). An opinion poll in Detroit found that, after Russia, 'what frightened people most was automation' ('The Coming of the Robots'). So while the committee itself was keen to emphasize the positive attitude towards automation from all sides of industry – 'not a single witness raised a voice in opposition to automation and advancing technology' (Joint Economic Committee 5–6) – the wider public framing of automation was not always as optimistic.

What is also interesting about the topic model in Table 1.1 is the way it highlights (e.g., in Topic 12, 'Definition of Word Automation'; Topic 19, 'Automation and Mechanisation') the controversy around the meaning and novelty of the word automation in the 1950s. One thing that becomes immediately apparent when we investigate the testimony referenced by these topics is that there is no single agreed meaning of automation. Indeed, the introduction to the report itself rather punts on the precise definition of automation: 'it is hardly a duty expected of a congressional committee to formulate, once and for all, a definition of a new word that is not yet in standard dictionaries' (Joint Economic Committee 4). In lieu of a single definition, the introduction instead offers three competing definitions: the first offered to them by Delmar Harder's colleague, Vice President of manufacturing at Ford, D.J. Davis: 'the automatic handling of parts between progressive production processes' (Joint Economic Committee 4, 53). This is essentially what Harder had been referring to when espousing the need for more automation: the creation of automatic transfer devices or mechanisms to move parts from one machining process to another. Alongside this, if you like, *Harderian* version of automation, the committee also referenced what we would now probably think of as the 'cybernetic' vision of automation: the use of electronic sensing for feedback and automatic control of machines and processes. Finally, there was the broadest version of automation, in which the new technologies would lead to a much more fundamental transformation of industry, one which had barely started and the significance of which, as the report's introduction states, 'we cannot predict and [one] with potentialities which we cannot fully appreciate' (Joint Economic Committee 4).

As is clear from the testimony about the industrial applications of automation, most of the actually existing automation in the mid-1950s was indeed of a Harderian kind. The problem for some committee witnesses was that Harderian automation was hardly different from existing production techniques: automatic transfer mechanisms were merely an extension of well-understood processes that had previously been referred to as mechanization. To put it another way, what was all the fuss about? Ralph J.

Cordiner, President of General Electric, suggested that 'you could say that progress toward greater automation is nothing new; only the expression "automation" is new' (Joint Economic Committee 425). Equally William W. Barton, President of W.F. & John Barnes Co., of Rockford, Illinois, a manufacturer of machine tools, believed that 'automation was a new word for a very old process', one that he traced back to the replacement of galley slaves by sail technology (Joint Economic Committee 245).[3]

While what I am calling Harderian automation remained mechanistic, there was clearly *something* novel about what Ford and other engineering companies were doing under the aegis of automation. One way of thinking about this is that they were implementing something which Larry Hirschhorn, in his brilliant 1986 book *Beyond Mechanization*, terms 'full mechanisation' (Hirschhorn 12). This term needs to be understood in the longer historical context of the transformation of American factories by electrification in the first part of the twentieth century. Nineteenth-century factories and mills usually had a single source of motive power, typically provided by a steam engine. This power was distributed to individual machines in the production process via drive shafts and belts. Such an arrangement created severe constraints on the positioning of machines and on their freedom of operation (for example, to speed up and slow down). Electric motive power, used to drive machines via small electric motors, allowed the reorganization and redesign of the factory, creating a more logical flow of parts around the factory (Hirschhorn 18–20). In this new context, automatic transfer machines could be designed to facilitate the flow of parts from process to process. Harderian automation, then, was the implementation of the dream of fully continuous, fully mechanized production.

But if Harderian automation represented merely an extension of mechanization, albeit of a potentially significant kind, other witnesses at the 1955 congressional hearings were convinced that automation represented a much more profound and novel kind of change. By far the most surefooted in terms of this use of the word automation was management consultant and automation evangelist John Diebold. Something of his confidence can be seen when, in the middle of a combative exchange on the kinds of investment needed for automation, Diebold corrected the language of fellow management and organization expert Walter Buckingham, whose own book, *Automation*, would be published in 1961:

Mr BUCKINGHAM. I think there is a difference in the use of the term 'investment' there. When a large corporation automizes – if that is a good term –

Mr DIEBOLD. Automates.

Mr BUCKINGHAM. Automates – it may invest in equipment or change its functions.... (Joint Economic Committee 38)

Here as elsewhere, there is a real sense in the Joint Economic Committee report of being an observer in the linguistic forge, while the words themselves are still red-hot. Diebold, once described by the *New York Times* as an 'evangelist of the future', was one of the pre-eminent writers and consultants on computerization in the postwar era (Bayot). His book *Automation* (1952) drove public recognition of the new term and helped to develop its meaning. It was no surprise, then, that when the US Congress held hearings on the automation issue in October 1955, Diebold was called as their first witness (Joint Economic Committee 6–29). In 1953, he wrote an article introducing the new technology for the *Harvard Business Review*. Diebold was keen to show that automation meant much more than simply advanced mechanization, the 'automatic loading and unloading of multistage machine tools', which Delmar Harder had in mind when coining the word (Diebold, 'Automation – the New Technology' 64). Automation, properly understood, was a whole new domain and required specialist expertise:

> When the electronics engineer and the computer designer write or talk of automation, however, they are referring to the application of a wholly new technology. To these people the word does not simply mean further mechanisation but rather symbolises a new technology which permits mechanisation of a very special sort. The concept of feedback is basic to this technology, and the idea of a closed-loop, self correcting system of controls is overriding. Oil refineries rather than automobile plants are used by these people as examples of the direction in which the automatic factory will develop — not merely because these process plants are more automatic but because they are designed around such a system. This represents quite a different meaning. The fact that it is not distinguished from the more general and less significantly new idea of automatic operation by authors and speakers on the subject of automation is unfortunate, though understandable at this early stage. For, unless the developments in this field are in fact understood as a new technology, there will be many mistakes made, and there will be many potential opportunities missed.
>
> (Diebold, 'Automation – the New Technology' 64)

Diebold's words exemplify several key dimensions of this early discourse on automation. Automation was different from mechanization, both in its operation and consequences for industry. It consisted of much more sophisticated systems of feedback and closed-loop control. It was to be driven by new electronic and computational technologies. Moreover, while elements could be described and understood in contemporary developments, full automation and its consequences were still only on the horizon. As we will see, automation is not only 'always already', but also always 'not yet'.

Diebold did not want to counter the threat of automation but to avoid missing the opportunities it afforded. He offered his business-minded readers a dramatic new outlook on industrial processes. Above all, automation had to be distinguished from mere mechanization. Diebold noted that Ford's supposedly '"automated" lines' in Cleveland and Buffalo included electric and hydraulic controls which permitted, 'automatic loading and unloading of multistage machine tools, while specially built "transfer" machines move work pieces from one operation to the next' (Diebold, 'Automation – the New Technology' 64). But such machines were 'nothing new ... only part of an old technology, developed a little further', an example of advanced mechanization, not automation. As Diebold argued, 'in approaching automation problems it is highly important that one does not think of automatic production as a process of automatically duplicating the hand motions of present machine operators' (Diebold, 'Automation – the New Technology' 66). Whereas mechanization simply substituted for the human hand, automation would be something different entirely. Nor did automation simply move from hand to brain, as Wiener had suggested. In fact, as Thomas Rid points out, Diebold was entirely opposed to the comparison Wiener was making between automatic control systems and the human brain (Rid 97). For Diebold, automatic control actually required changing the whole way in which a product was made, rather than the replacement of human brains with computers. The real difference was that whereas mechanization simply meant taking a particular human action on the assembly line and substituting it with a machine, automation required the complete structural reorganization of the production process.

Redesigning the production process might equally mean redesigning the product. Diebold gives the example of the British company Sargrove Electronics who had come up with an automatic machine called ECME (Electronic Circuit Making Equipment), with the aim of building a complete radio set every 20 seconds (Popular Science, 'Robot Squirts Out 3 a Minute' 160). Sargrove replaced the then-traditional wiring chassis with plastic plates. By spraying on a conductor and then machining the output, ECME automatically built up capacitors, resistors and potentiometer tracks. Crucially, from Diebold's perspective, this allowed for modular redesign of the production process as the product changed (1953 67). Sargrove's process was not only far ahead of existing techniques for manufacturing electronic circuits but arguably more advanced than the printed circuit board technique that eventually saw widespread adoption from the mid-1950s onward. PCBs only provided conduction tracks linking together discrete components that had to be manually inserted and soldered onto the board, whereas ECME allowed simple components to be moulded and sprayed onto the board directly. Yet Sargrove's radio sets never achieved large-scale production, and the automated production line that had achieved widespread acclaim in publications such as *Popular Science* may well have functioned primarily as a demonstrator (UK Vintage Radio Forum). Sargrove was a poster child for automated production that never actually came to exist.

Nonetheless, Sargrove was for Diebold a promising example of how automation, unlike mechanization, meant completely rethinking from the ground up both the manufacturing process and the product. But one might doubt, both as a matter of principle and as a matter of fact, the distinction Diebold seeks to make between automation, on the one hand, and advanced mechanization on the other. Is it clear, for example, that all that is involved in mechanization is just the substitution of machine for human hand? For an alternative account, one might look at Siegfried Giedion's magisterial 1948 tome *Mechanization Takes Command*. Giedion emphasized the role that mechanization plays in the structural transformation of whole industries. In bread production, for example, mechanization changes the whole organization of production from milling to baking. But it also required a dramatic rethinking of the product. For example, flour needed to be bleached rather than aged in order to reduce warehousing costs (Giedion 189). Ultimately, mechanization resulted in a transformation of public taste in bread. So over time and across whole industries, Giedion's version of mechanization can be seen as systematically transforming both process and product in a similar manner to Diebold's version of automation.

However, Dieboldian automation also supposedly differed from mechanization in its embrace of the concepts of feedback and control. Here, Diebold took on board Norbert Wiener's popular and ground-breaking account of cybernetics, but also distanced himself from it in important ways. Wiener argued that new automatic control systems emulated living organisms in their ability to take account of feedback, for example, from sensors, and used that to modify the action they were designed to regulate. Deployment of these systems would result in machines that could not only take an action, but could also respond to its actual rather than intentional results (Rid; Wiener 26–7). Wiener illustrated this point with well-known existing mechanical feedback systems. The domestic thermostat responded to sensed temperature (in traditional thermostats, via a bimetallic strip that bends in response to heat) by regulating the heating system for a house; James Watt's centrifugal governor regulated the speed of a steam engine by closing a valve as it spun faster and opening it again when speed decreased. Wiener's interest in feedback stemmed from his involvement in US wartime efforts to build feedback-based systems to control anti-aircraft fire by predicting the paths of aircraft (Kline 20–5). Both Wiener and Diebold saw that in the postwar era, these ideas could have widespread civil application. But as Diebold testified to Congress, it was not so much the principle itself – a self-regulating control system – as the novel ways in which it was to be *applied* that would shape the new technology, 'properly called automation' (Joint Economic Committee 9). While feedback and control were clearly important components of what he described, Diebold was at pains to make clear that automation was not reducible to the cybernetic logic of control. He argued:

> Paradoxically, the current obsession with the novelty and spectacular performance of automatic controls diverts attention from the problems of their application to industry. Although automatic control mechanisms are *necessary* for the achievement of fully automatic factories, they are not *sufficient* in and of themselves. The full promise of this technology cannot be realised as long as we think solely in terms of control.
>
> (Diebold, *Automation* 2)

Necessary, but not sufficient, then, 'control' was also a form of 'loose thinking'. As concepts, automatic control and the electronic computer had, for Diebold, captured the public imagination in an unhelpful manner. The relationship between the 'operation of military equipment' (Diebold is probably thinking here of Wiener's automatic fire-control systems) and changes in the industrial machinery of the 1950s was, at best, superficial (Diebold, *Automation* 3–5). What was needed was a much more holistic thinking about the organization of production and the use of new technologies. When asked in 1955 to give his own definition of the word automation, Diebold described it as 'a means of analyzing, organizing, and controlling our production processes to achieve optimum use of all our productive resources – mechanical and material as well as human' (Joint Economic Committee 15).

Dieboldian automation, then, represented a much more profound industrial, economic and even social change. His is perhaps the most convincing – and yet pragmatic – articulation of the idea that automation was a novel force and an important shift within the previous history of mechanization.

As I argued in the introduction to this chapter, contemporary accounts of the discourse around automation in the 1950s tend to view it exclusively through the lens of cybernetics. But both Harderian and Dieboldian automation show that there was actually a variety of discursive forces at work in the 1950s debates, and the 'cybernetic' ideas are only one element among them.

The Automation Debates

In recent work, Caroline Bassett and I have theorized the recurrent debates about automation in the postwar era as 'automation anxiety'. We examine ongoing concerns about (and enthusiasm for) automation, tracing this history from cybernetics to contemporary concerns about artificial intelligence (Bassett and Roberts, 'Automation Anxiety'). We argue that these anxieties, while seemingly triggered by technological advancements, are rooted in deeper sociopolitical issues surrounding labour, work and the future of society. We propose a critical historical approach to understanding the cyclical nature of automation anxiety, highlighting the enduring relevance of past debates in shaping current discussions about AI and its societal impact. We argue against the 'technological solutionism' of some work on 'responsible AI' and its replacement by more nuanced perspectives that consider the complex interplay between technological change and broader social anxieties. We also look at different contemporary political responses to automation anxiety, including accelerationism and calls for universal basic income. We investigate how automation debates have influenced contemporary left-wing thought, particularly the work of Nick Srnicek and Alex Williams and that of Aaron Bastani, which envision emerging AI-based automation as a potential pathway to a post-capitalist utopia. By comparing past debates with current discussions around post-capitalism, we contest their novelty and argue for a productive critical confrontation of the present by the past (Bassett and Roberts, 'Automation Now and Then').

The etymology of the word automation discussed in my opening section curiously foreshadows these cyclical debates about automation that have waxed and waned from the immediate postwar era until the present day. Automation is always new and also not new. Every fresh wave of automation anxiety has its seers who prophesy the imminent arrival of an unprecedented new age of machines, challenging the very foundations of industry, employment, government, society and even humanity. Every fresh wave of automation anxiety equally has its doubters: those who question the extravagant promises made for new technology and query both the urgency and novelty of its claims on our attention. As Bassett and I have argued, the cyclical nature of these automation debates, their not-newness, ought to make us cautious about the idea that they result from the latest technical developments: advanced mechanization, closed-loop control systems, increased computation, machine learning and so on.

The influential 1978 BBC documentary 'Now The Chips Are Down' illustrates well, in a British context, the way in which debates about automation from the 1950s and 1960s return in response to later technical developments, specifically in this case the rise of Silicon Valley and the emergence of cheap microprocessors. As in previous waves of automation anxiety, the documentary foresaw an imminent qualitative shift in the role automation would play ('Automation is coming in every farm, office, factory and shop') and revived key ideas from previous debates, suggesting that automation might lead to large-scale unemployment and increasing gaps between rich and poor.

Presciently, the documentary also warned that unions would resist automation, something that later became evident in the 1980s battles where print unions resisted the computerization of newspaper production, such as the Wapping Dispute (1986).[4] The documentary attracted considerable attention and was even screened to then-Prime Minister James Callaghan's cabinet (Ferguson 68; Lean 547). The reaction to 'Now The Chips Are Down' was such that the BBC's Continuing Education Department was invited to write a report for the UK government's Manpower Services Commission (Albury and Allen) on how the anticipated crisis could be responded to through education and training. This later led to the creation of the BBC's own Computer Literacy Project and the creation of the iconic BBC microcomputer or 'BBC Micro'. By 1985, the BBC Micro was in use by 75% of schools in the UK, helping to nurture a new generation of programmers.

The BBC Micro was a large commercial success for the small British company chosen by the BBC to develop it, Acorn Computers. In 1987, Acorn released a more advanced computer, based on a RISC (Reduced Instruction Set Computer) CPU they had developed in-house (Gazzard 168–9). The part of the company involved in designing the chip was later spun off into its own company, Advanced RISC Machines, or ARM for short, still based in Cambridge. As Alison Gazzard notes, by 2007, ARM processors featured in 98% of the more than 1 billion phones sold worldwide (170). Today in 2025, virtually every Android and every Apple iOS-based device features a processor with either ARM-designed cores or using ARM's architecture.

'Now The Chips Are Down' is, therefore, an exemplary case study in the relationship between cultural debates and technological development. 1970s British automation anxiety, far from being simply a reaction to developments in microchip technology and the success of Silicon Valley, helped to shape policy responses and boost the UK's emergent microcomputer industry, eventually catalyzing the growth of commercial semiconductor design in the UK. It also demonstrates, as Bassett and I have argued, that cyclical automation debates are not directly anchored in particular technological developments but are part of a broader sociotechnical envisioning or technological imaginary. Automation technologies and debates run in parallel and influence each other. Moreover, the 1970s debates in the UK drew on a history of thinking about automation going back to the 1950s and beyond, just as our own debates about AI-based automation reanimate those of previous eras.

In part, this becomes apparent by examining the types of futures that were imagined in previous debates about automation, and the fate of those futures. In relation to the visionaries and evangelists of the 1950s, Harderian, Dieboldian and cybernetic models of automation all ultimately proved to have severe limitations. As Hirschorn shows, the model of continuous production through automated transfer devices proved to be less useful than imagined, in part because the design of such machines was specific to the assembly of a particular product; changes in demand and new models of products required costly re-engineering. Hirschorn notes wryly, 'the full mechanisation of the supermachine could operate only with the standardisation of demand and a long-term plan for coordinating buyers' wishes with sellers' capacities' (Hirschhorn 13). Harderian automation was real but rigid and not a replacement for the flexibility of human workers. Indeed, economic fears about the impact of automation on employment had already started to recede by the mid-1960s when Daniel Bell, for example, published an article entitled 'The Bogey of Automation'.

Meanwhile, by the 1980s, the cybernetic vision of the self-regulating machine and electronic feedback and control was also proving less convincing. Events such as the Three Mile Island nuclear disaster in 1979 had provided stark examples of what could happen when control systems encountered conditions outside their design. For Hirschorn, the implications of such accidents were that new cybernetic machines required more skilled workers who were able to understand the new errors the control systems encountered and respond to them. He contended that, 'the cybernetic image of the perfect machine is as utopian as was its nineteenth-century counterpart' (Hirschhorn 1). Where Harderian automation introduced new rigidity, cybernetic automation had resulted, against its proponents' expectations, in a new fragility.

Examining the emergence of the term automation in the postwar era demonstrates the significance and subtleties of automation debates in this period. The tendency to reduce automation to a generic, ahistorical generalization often downplays the importance of its emergence as a concept (albeit, slippery) in the postwar era. Equally, understanding automation in this postwar period exclusively through the framework of cybernetics fails to do justice to the complexity of the debates in that era.

Notes

1 See also on this (Bassett and Roberts, 'Automation Anxiety').
2 The report text is available from https://www.jec.senate.gov/public/index.cfm/1956/12/report-c2e5856e-06c9-497c-93bc-65765bb22a98. The text was extracted from the PDF using pdftotext and broken into approximately paragraph-length chunks using GNU awk. Topic modelling was performed with BERTopic using a 'Count Vectoriser' model provided by scikit-learn (Noonburg; *Gawk*; Grootendorst, *BERTopic*; Pedregosa et al.)
3 As Sebastian Franklin has discussed in recent work, the language and metaphor of slavery, such as the concept of 'mechanical slaves', is a constant presence in early discussions of automation (Franklin).

4 Caroline Bassett's work points to the ways in which this and similar print industry disputes belong to the longer history of what she calls 'anti-computing' (Bassett 12).

References

Adams, Paul G. "A Discussion of Automation." Thesis. George Washington University, Washington, District of Columbia, 1956, https://calhoun.nps.edu/handle/10945/31036.

Albury, Robert, and David Allen. *Microelectronics*. BBC Continuing Education Department, Dec. 1979, http://www.naec.org.uk/organisations/bbc-computer-literacy-project.

Bassett, Caroline. *Anti-Computing: Dissent and the Machine*. Manchester UP, 2022, https://library.oapen.org/handle/20.500.12657/53144.

Bassett, Caroline, and Ben Roberts. "Automation Anxiety: A Critical History – the Apparently Odd Recurrence of Debates about Computation, AI and Labour." *Handbook of Critical Studies of Artificial Intelligence,* edited by Simon Lindgren, Elgar, 2023, pp. 79–93, https://www.elgaronline.com/edcollchap/book/9781803928562/book-part-9781803928562-12.xml.

Bassett, Caroline, and Ben Roberts. "Automation Now and Then: Automation Fevers, Anxieties and Utopias." *New Formations*, vol. 98, 2019, pp. 9–28, https://doi.org/10.3898/NEWF:98.02.2019

Bastani, Aaron. *Fully Automated Luxury Communism: A Manifesto*. Verso, 2019.

Bayot, Jennifer. "John Diebold, 79, a Visionary of the Computer Age, Dies." *The New York Times*, 27 Dec. 2005, https://www.nytimes.com/2005/12/27/nyregion/john-diebold-79-a-visionary-of-the-computer-age-dies.html.

Bell, Daniel. "The Bogey of Automation." *The New York Review of Books*, vol. 5, no. 2, Aug. 1965, pp. 23–5.

Benanav, Aaron. *Automation and the Future of Work*. Verso, 2020.

Brett, Megan R. "Topic Modeling: A Basic Introduction." *Journal of Digital Humanities*, vol. 2, no. 1, 2012, http://journalofdigitalhumanities.org/2-1/topic-modeling-a-basic-introduction-by-megan-r-brett/.

Department of Scientific and Industrial Research. *Automation: Technical Trends and Their Impact on Management and Labour*. Department of Scientific and Industrial Research, Great Britain, 1956.

Diebold, John. *Automation: The Advent of the Automatic Factory*. Van Nostrand, 1952.

Diebold, John. "Automation – the New Technology." *Harvard Business Review*, vol. 31, no. 6, Dec. 1953, pp. 63–71.

Diebold, John. "Automation Needs a Human Policy." *Challenge*, vol. 7, no. 8, 1959, pp. 42–6.

Diebold, John. "Life Under Automation." *Challenge*, vol. 9, no. 3, 1960, pp. 20–5.

"Fairless Upholds 'Automation' Aims: U. S. Steel Chairman Asserts Fears of Joblessness Are 'Vicious Propaganda'." *New York Times*, 12 Feb. 1955, p. 27.

Ferguson, Majorie. "The Challenge of New-Technological Determinism for Communication Systems, Industry and Culture." *New Communication Technologies and the Public Interest: Comparative Perspectives on Policy and Research*, edited by Marjorie Ferguson, Sage, 1986, pp. 52–79.

Finney, Burnham. "Origin of 'Automation'." *The New York Times*, 11 Apr. 1965, p. 168.

Franklin, Sebastian. *Mechanical Slavery: Freedom, Subjection, Automation.* Stanford UP, 2026.

Freeman, William M. "Automation Aims at New Freedom: Devices That Run Factories Promise to Release Men for Richer Living." *The New York Times*, 3 Jan. 1955, p. 45.
Gawk. 5.0.1, Free Software Foundation, https://www.gnu.org/software/gawk/. Accessed 7 Feb. 2025.
Gazzard, Alison. *Now the Chips Are Down: The BBC Micro*. MIT Press, 2016.
Giedion, Sigfried. *Mechanization Takes Command: A Contribution to Anonymous History*. U Minnesota P, 2013.
Grootendorst, Maarten. "BERTopic." 2020. 0.16.4, https://github.com/MaartenGr/BERTopic. Accessed 7 Feb. 2025.
Grootendorst, Maarten. *BERTopic: Neural Topic Modeling with a Class-Based TF-IDF Procedure*. arXiv:2203.05794, arXiv, 11 Mar. 2022, https://doi.org/10.48550/arXiv.2203.05794.
Hirschhorn, Larry. *Beyond Mechanization: Work and Technology in a Postindustrial Age*. MIT Press, 1990.
Joint Economic Committee. *Automation and Technological Change*. Hearings (84th Congress), United States Congress, 1956, https://www.jec.senate.gov/public/index.cfm/1956/12/report-c2e5856e-06c9-497c-93bc-65765bb22a98. Accessed 7 Feb. 2025.
Kline, Ronald R. *The Cybernetics Moment: Or Why We Call Our Age the Information Age*. Johns Hopkins UP, 2017.
Lean, Thomas. "Mediating the Microcomputer: The Educational Character of the 1980s British Popular Computing Boom." *Public Understanding of Science*, vol. 22, no. 5, Jul. 2013, pp. 546–58, https://doi.org/10.1177/0963662512457904.
Marshall, Margaret Wiley. "'Automation' Today and in 1662." *American Speech*, vol. 32, no. 2, 1957, pp. 149–51, https://doi.org/10.2307/453032.
Marshall, Margaret Wiley, and R. W. Burchfield. "'Automation' Today and in 1662." *American Speech*, vol. 34, no. 3, 1959, pp. 236–7, https://doi.org/10.2307/454058.
Moretti, Franco. *Distant Reading*. Verso, 2013.
Noonburg, Derek. *Pdftotext*. 3.03, Glyph & Cog, LLC, https://www.xpdfreader.com/pdftotext-man.html. Accessed 7 Feb. 2025.
Oxford English Dictionary. *Automaton, n., Etymology*. Oxford UP, Dec. 2024, https://doi.org/10.1093/OED/8972522451.
Pedregosa, Fabian, et al. "Scikit-Learn: Machine Learning in Python 2010." *Journal of Machine Learning Research*, 1.6.0, vol. 12, 2011, pp. 2825–30, https://jmlr.csail.mit.edu/papers/v12/pedregosa11a.html.
Pollock, Friedrich. *Automation: A Study of Its Economic and Social Consequences*. Translated by W. O. Henderson and W. G. Chaloner, Praeger, 1957.
Rid, Thomas. *Rise of the Machines: The Lost History of Cybernetics*. Scribe, 2016.
"Robot Squirts Out 3 a Minute." *Popular Science*, vol. 152, no. 4, Apr. 1948, pp. 160–1.
"Special Report: Automation." *Business Week*, vol. 1361, 1 Oct. 1955, pp. 74–102.
Srnicek, Nick, and Alex Williams. *Inventing the Future: Postcapitalism and a World Without Work*. Verso, 2015.
"The Coming of the Robots." *The Economist*, vol. 177, 5 Nov. 1955, p. 481.
UK Vintage Radio Forum. "Sargrove. Type A3 Radio." *UK Vintage Radio Repair and Restoration Discussion Forum*, 2012, https://www.vintage-radio.net/forum/showthread.php?t=88980. Accessed 7 Feb. 2025.
Wells, H. G. *The Time Machine*. Pocket Books, 2004.
Wiener, Norbert. *The Human Use of Human Beings: Cybernetics and Society*. 1954. Free Association, 1989.

2 When the Clock Took the Floor

Technology as Non-Human Actor in Augusto De Angelis' Detective Novel *Il Banchiere Assassinato* (1935)

Emanuele Stefanori

Detective fiction is commonly known in Italy as *giallo* (yellow), a name that originates from the yellow covers of the first book series dedicated entirely to the genre, *I libri gialli*, launched by the Milanese publisher Mondadori in 1929. This date marks a symbolic starting point for the circulation of detective stories in Italy, which emerged during the Fascist era. The sociopolitical environment of the time significantly influenced the genre's birth and development in the ensuing years. Despite the increasing popularity of detective stories, Italian authors and publishers soon had to contend with the limitations and dictates of Fascist censorship. Specifically, by addressing crimes and murders, the Italian *giallo* undermined the Fascist rhetoric of a crime-free, safe country – one of the key points of the regime's propaganda (Crovi, *Storia* 68). Moreover, it cast the official police in a bad light, portraying them as inept at solving cases and often ridiculed by the exceptional intelligence of private investigators, in line with the conventions of the genre (Pistelli 234). Despite restrictive laws aimed at limiting or regulating the publication of crime fiction during the Fascist era, many Italian writers continued to publish detective novels, attempting to evade or circumvent these forms of censorship as much as possible (Pezzotti 17). This sociopolitical context led the Italian *giallo* to forge its own identity and set of conventions, moving away almost immediately from classic detective fiction. The Italian *giallo*, therefore, was subversive and politically engaged from the outset, even if only implicitly, by the mere fact of having challenged Fascist censorship. The notion of political engagement, which would come to characterize the genre in later years, was already present in the works of the earliest authors, albeit subtly. In a way, due to the regime's aversion to British and North American cultural models, the Italian *giallo* was almost forced to find an alternative path from the very beginning.

Augusto De Angelis was by far the best detective fiction writer of the first generation, due to both the quality of his writing and his literary commitment to the crime genre. Born in Rome in 1888, De Angelis was a

DOI: 10.4324/9781003545910-3

novelist, journalist and playwright. He published almost 20 detective stories between 1935 and 1943, featuring the main character, Commissioner Carlo De Vincenzi of the Milanese *squadra mobile* (flying squad).[1] He also wrote an important short essay on detective fiction in the late 1930s. During the Second World War, De Angelis was accused of being an anti-Fascist activist and arrested in 1943, although he was released from prison a few months later. However, he died shortly after at Como Hospital in 1944, following a violent assault by a Blackshirt in Bellagio. Hidden between the lines of apparently simple detective stories, De Angelis' 'multi-layered anti-fascism' (Pezzotti 31) is evident in his novels as well as his passionate defence of the genre, which can also be seen as 'an attack on the regime's interference in Italian cultural production' (25).

In this chapter, I analyse De Angelis' debut detective novel, *Il banchiere assassinato* (*The Murdered Banker*), as a perfect example of the original approach to the genre, which contributed to the creation and evolution of a distinctively Italian detective model. I will focus particularly on the complex relationship between the policeman and technology to better define the very characteristics of the Italian fictional sleuth. Specifically, I will explore Commissioner De Vincenzi's conflicted relationship with technology and modern investigative methods, characterized by his fascination with mechanical objects – an influence of the Futurist glorification of the machine – as opposed to a psychological approach to investigation, inspired by Freud's psychoanalysis and G.K. Chesterton's Father Brown mysteries. By characterizing himself as a poet-detective, Commissioner De Vincenzi openly rejects the positivist approach of the classic detective novel, exemplified by the analytical coldness and flawless rationality of exceptional investigators like Sherlock Holmes.

Detective fiction can be seen as an ideal point of convergence between science and literature due to the crucial role of technology and forensic science in the narrative structure. Through a close reading of the novel, technology and non-human agency in De Angelis' crime story will be examined on four distinct yet complementary levels. First, the preference for a psychological approach to investigation reflecting the creation of a distinctly Italian type of detective; second, the narrative process of humanizing technological and mechanical objects while dehumanizing human characters, portraying them as automaton-like figures in a mechanized cityscape; third, the important role of technological devices as non-human actors in the story, exemplified by a pendulum clock in a crucial moment of the investigation; and finally, the detective's nuanced relationship with forensic science and modern methods of investigation.

Some key concepts from Bruno Latour's Actor-Network Theory can help analyse the agency of technology in the story. According to this theoretical approach, technological objects are embedded in social life and are placed on the same level as human actors in the process of social construction. Actor-Network Theory terms actor – or actant – anything that can alter or change the state of things by making a difference, including human beings, non-human animals, technological devices, physical objects, and even concepts or ideas. The principle of generalized symmetry is adopted, treating human and non-human actors as equally important. This means that non-human elements like technologies or scientific instruments are as significant as human actors in creating networks. Depending on the nature of their agency, actors can also be defined as either mediators or intermediaries. Specifically, actors are considered as mediators when they are able to transform, translate, and modify meaning and outcomes, or as simple intermediaries when they transmit information unchanged. Both human and non-human actors are part of networks, which are dynamic and constantly changing social aggregates, in which actors interact and influence one another. Therefore, when analysed through the lens of Actor-Network Theory, human and non-human actors have the same importance in the narrative construction of the social context in which the criminal event takes place. They similarly enjoy equal importance in the process of shaping the social groups over the course of the investigation, eventually coming to a reconfiguration of these groups with the resolution of the case. According to Latour, 'social aggregates are not the object of an ostensive definition—like mugs and cats and chairs that can be pointed at by the index finger—but only of a performative definition' (34). In other words, social groups are dynamic and constantly transforming. In Latour's concept of 'sociology of associations' (9), the social is not a pre-established condition but is actively constructed and continuously shaped and reshaped through practices, actions, and interactions between human and non-human actors. Combining such a 'performative definition' of social aggregates with Ernest Mandel's idea of detective fiction as the representation of 'bourgeois ideology *par excellence*' (Mandel 47), technology can be read as involved – along with human characters – in a dynamic process of redefining social groups, which results in the 'performative' assemblage of human and non-human actors.

Augusto De Angelis was not only one of the founding fathers of the Italian *giallo* and 'the finest Italian detective novelist of his generation' (Dunnett 16). As Luca Somigli points out, he was also the first Italian author to theorize a specifically Italian approach to detective fiction with a short essay written in the late 1930s. Part of this text was used as an introduction to the novel *Le sette picche doppiate* (1940), but the essay was only republished in its entirety only in the 1980s with the posthumous title *Conferenza sul giallo [in tempi neri]*.[2] De Angelis' defence of the Italian *giallo* is an affirmation of this genre's ability to provide 'a critical representation of the contradictions of modernity' (Somigli 71), especially in the difficult period of the Fascist era. According to De Angelis, the experience of modernity leads to the development of this new literary genre. More precisely, the experience of modernity creates a series of social anxieties that detective fiction is able to represent but also to dispel.

Il banchiere assassinato (1935)[3] is De Angelis' debut detective novel, featuring Commissioner De Vincenzi as the protagonist. One evening in 1930s Milan, Commissioner De Vincenzi is visited by his old university friend, Giannetto Aurigi. That same night, the banker Mario Garlini is found murdered in Aurigi's apartment. De Vincenzi soon discovers that Aurigi had significant debts with Garlini's bank. Furthermore, the discovery of a recently fired gun and a vial of prussic acid (hydrogen cyanide) in Aurigi's apartment casts further suspicion on De Vincenzi's friend. As the investigation unfolds, De Vincenzi also begins to suspect Aurigi's fiancée, Maria Giovanna, and her father, Count Marchionni, as both have motives to kill the banker. Nevertheless, De Vincenzi's careful reconstruction of events ultimately reveals Aurigi's butler as the true culprit. Put simply, the butler did it!

Despite the clichéd story, *Il banchiere assassinato* is a cornerstone of the Italian *giallo*. The novel opens with a glimpse of the city of Milan late at night. The central Piazza San Fedele is wrapped in a thick fog while people are walking across the square:

> Piazza San Fedele era un lago bituminoso di nebbia, dentro cui le lampade ad arco aprivano aloni rossastri. L'ultima auto si allontanava lentissimamente dal marciapiede del teatro Manzoni, facendo risuonare sordamente il clacson. Il teatro chiudeva le sue grandi porte nere. Qualche ombra fantomatica attraversava la piazza. Due ombre si scontrarono sullo sbocco di Via Agnello e una di esse notò che l'altra era quella di un signore in abito da sera, pelliccia e tuba. Il signore per suo conto non vide che un'ombra nera. Non guardava neppure, del resto. Camminava. Procedette dalla piazza per Via Agnello, nella nebbia, lentamente. Andava.
>
> (*BA* 5)

> [Piazza San Fedele was a bituminous lake of fog penetrated only by the rosy haloes of arched street lamps. With a muffled sound of the horn, the last car moved slowly away from the pavement outside the Manzoni Theatre. The theatre closed its great black door. Ghostlike shadows were moving across the piazza. Two of them collided at the opening of via Agnello and one noticed that the other was that of a gentleman in evening dress, fur coat and top hat. For his part, the gentleman saw nothing but a black shadow. In any case, he wasn't looking. He was walking. From the piazza he proceeded slowly through the fog along via Agnello, continuing on his way.]
>
> (*MB* 7)

De Angelis sets the crime story in a modern (and modernist) Milan. The choice of the city elements presented in this opening paragraph is particularly significant. He starts with electrical and mechanical elements ('le lampade ad arco', 'l'ultima auto', 'il clacson'), namely technological elements, before introducing human characters. However, they are not described as individuals walking on the streets but presented instead as dark silhouettes akin to shadows or ghosts. One of these ghost-like shadows moves almost mechanically without even noticing the other individual he is running into, much like an automaton. In addition, phrasing and punctuation in this paragraph highlight the idea of mechanical movement by isolating the two verbs of motion ('camminava', 'andava') in one-word sentences.

The description of Milan with its streets and buildings shrouded in dense mist places it in the Gothic register – the perfect noir environment for a crime story – but it is also a modern urban space shaped by technological elements. As Somigli notes, 'De Angelis exalts modern technology in terms that closely recall the futurist peans of the machine' (Somigli 75). The description of Milan in Futurist terms clearly echoes passages from De Angelis' essay on detective fiction:

> La Vita ha preso il sopravvento – dissennata iconoclasta dalle mani d'acciaio – e tutto rompe e scuote, tutti i sogni sovverte e al posto delle nuvole mette i grandi cumuli di scorie metalliche, che forni incandescenti vomitano; sopra i vialetti di narcisi, di rododendri e di clematidi fa scorrere la lava livellatrice dei suoi asfalti; abbatte i templi ed eleva i grattacieli; spezza le gambe ai bei ginnetti dall'inanellato crine e lancia tra i viali di citisi e di larici le sue macchine aerodinamiche, che tagliano le fronde e scortecciano i tronchi col rasoio dei loro parafanghi.
>
> (*Conferenza* 44)

> [Life has taken over – a senseless iconoclast with steel hands – and it breaks and shakes everything, subverts all dreams, and instead of clouds, it places great heaps of metallic waste, which incandescent furnaces vomit up; over the paths of daffodils, rhododendrons, and clematis flows the levelling lava of its asphalts; it demolishes temples and erects skyscrapers; it breaks the legs of beautiful ponies with ringleted manes and sends its aerodynamic machines down the avenues of laburnums and larches, cutting the branches and stripping the bark from the trunks with the razor of their fenders.]
>
> (My translation)

Many references to Futurism are evident in this paragraph, as life is closely linked to technology, which fundamentally alters the conditions of human experience and 'reshapes the landscape of modernity' (Somigli 75).

One of the individuals in the fog is Commissioner De Vincenzi, heading towards his office in Piazza San Fedele. Once at the police station, De Vincenzi meets the journalist De Blasi. The novel's opening dialogue, therefore, occurs between a detective and a journalist, symbolically representing two fundamental components of the narrative structure due to the crucial role of news circulation in detective fiction. While introducing a key element of the crime genre, alongside the narrative trope of the love-hate relationship between the police and the mass media, De Angelis uses a technological metaphor to depict the chubby, ruddy-faced journalist. The reader learns that De Vincenzi likes De Blasi, 'quel giornalista, tondo e rosso come un segnale di via ingombra' (*BA* 6) [This journalist, round and red as a traffic signal at a blocked junction (*MB* 8)]. The metaphor of the traffic signal points again to modernity, urban space and the mechanization of life. On one hand, it places the journalist within the cityscape, as part of the urban environment, metonymically emphasizing the strong relationship between the city and mass media. On the other hand, the technological metaphor partially dehumanizes the reporter, following the description of the automaton-like individual walking in the fog. In other words, both men are components of the technological and mechanized urban space.

The narrative mechanization of these characters, depicted as cogs in a technological cityscape, serves to emphasize De Vincenzi's human traits. The author portrays the detective not only as human but as a humanist-sleuth, creating a new kind of detective in De Vincenzi – a poet-detective. Both the character and his author consider themselves poets: for De Vincenzi, working as a police commissioner is a way to express his poetic sensibility; similarly, for De Angelis, writing detective stories is like composing verses (Rambelli, *Storia* 97). He uses his love of poetry to defend the detective-fiction genre and almost provocatively states that 'nulla è più poetico di un romanzo poliziesco!' (*Conferenza* 29) [Nothing is more poetic than a detective novel! (My translation)]. According to De Angelis, detective fiction captivates modern readers for its reassuring power and ability to dispel new and irrational fears generated by the hectic lifestyle of modern cities. Yet, as Brunetti suggests, readers can also find a melancholic poem secretly embedded in the crime genre (107). This kind of poetry is defined by the frantic life of the present, so the poetry hidden in detective stories is the same as that found in modern life:

> Anco le stelle subiscono le vibrazioni dei motori e sul cielo azzurro, la fanciulla dal cuore puro non vedrà più volo d'angeli e non leggerà più rime tenere, ma gli annunci reclamistici dell'ultimo match o della più veloce automobile tracciati a pennellate effimere dalla coda degli aeroplani. No, (...) non ho rimorsi a scrivere oggi romanzi polizieschi. È un far versi anche questo. Anche al problema e all'enigma occorrono

le rime. E se il rimario di cui ci si serve è il manuale di criminologia o il trattato di Tardieu sui sintomi e sul decorso dei veleni, siamo sempre poeti. Anche se le api d'oro tacciono spaurite sull'Imetto pieno di mirti.
(*Conferenza* 44)

[Even the stars feel the vibrations of the engines, and in the blue sky, the pure-hearted girl will no longer see the flight of angels or read tender rhymes, but the advertisements for the latest game or the fastest car, traced in ephemeral strokes by the tails of airplanes. No, I have no regrets about writing detective novels today. This, too, is like writing poetry. Even problems and enigmas need their rhymes. And if the rhyming dictionary we use is a criminology manual or Tardieu's essay on the symptoms and progression of poisons, we are still poets. Even if the golden bees remain silent and frightened on a myrtle-covered Hymettus.]
(My translation)

The detective story is the most genuine new form of poetry, placing the idea of death at the very centre of everything, amidst the chaotic present and restlessness of life. However, the poet's aim is not to use poetry to give new life to what is dead; instead, they will use the new 'rhyming dictionary' only to explain the causes of death (Brunetti 110).

The fear of death, also present in the novel, reflects Freud's influence and leads to the psychological approach to investigation. As De Angelis clearly states, 'oggi per scrivere un romanzo poliziesco, occorre conoscere Freud' (*Conferenza* 33) [Today, to write a detective novel, one must know Freud (My translation)]. In the classic whodunit, the detective novel contains two stories: one story consists of the murder and the other of the investigation, so 'the first story, that of the crime, ends before the second begins' (Todorov 44). In De Angelis' detective novels, however, the presence of the dead body serves as the magnetic pole of attraction of *both* stories in a deeper way. For at the very centre of his stories is not only the fear of death but also fear of the dead body itself. By emphasizing the corporeality of murder, De Angelis' *gialli* are also a perfect example of detective fiction as the expression of bourgeois fear of death. As Ernest Mandel points out, fear of death in bourgeois society is not just fear of something natural and unavoidable for human beings. Bourgeois obsession with death as an accident leads to the idea of violent death. In detective fiction, death is no longer treated as an ontological question or individual tragedy but as an object of investigation. The preoccupation with human destiny is replaced by the preoccupation with crime, so there is only 'a corpse to be dissected, a thing to be analyzed. Reification of death is at the very heart of the crime story' (Mandel 41). In De Angelis' novels, 'il cadavere è presente in ogni pagina, calato in un mistero profondo e greve. È per non sapere reggere a questa presenza ineffabile che l'assassino si abbatte stroncato' (Rambelli, *Storia* 103) [The dead body is present on every page, shrouded in a profound and heavy mystery. The murderer collapses, overwhelmed, because they cannot bear this ineffable presence (My translation)]. Much like a Freudian nightmare, the investigation functions as a psychoanalytic process aimed at purging the presence of death from the characters' subconscious, while the lingering psychological stress fosters a deeply unsettling atmosphere.

De Vincenzi's psychological analysis of people and events often leads to a symbolic reconstruction of reality. Even the concept of time is occasionally altered, and the surrounding physical environment is no longer described in realistic terms (Rambelli, *Acculturazione* 112). When this occurs, objects and technologies also appear transformed. In the first chapter, the telephone on De Vincenzi's desk is more than a simple mechanical tool. While the individuals in the fog and the reporter in the office undergo a process of narrative mechanization, the telephone, in De Vincenzi's eyes, seems to take on human characteristics, becoming like an old friend or even a lover.

> De Vincenzi mise la mano sulla scatola nera e lucida, toccandola quasi amorosamente.
>
> «Il mio caro, tirannico telefono! È lui che nella notte, nelle lunghe ore di veglia, mi unisce alla città ... Esagero. Diciamo al mondo, al mio mondo di commissario, capo della Squadra Mobile [....] Guardalo ...! È tozzo, nero, inespressivo per te. Nient'altro che una scatola con una buffa cornetta e un cordone verde. Ma per me ha mille voci, mille volti, mille espressioni. Quando squilla, io so già se mi reca un richiamo d'ordinaria amministrazione oppure se mi annuncia un nuovo dramma, una tragedia d'amore e di delinquenza...».
>
> (*BA* 9)

> [De Vincenzi put his hand on the shiny black box, touching it almost lovingly. 'My dear, tyrannic telephone! It's this which, during the long hours of waiting at night, connects me to the city ... I'm exaggerating. Let's say to the world, my world as a policeman, head of the flying squad (....) Look at it! It's chunky, black, inexpressive. For you, nothing but a black box with a silly mouthpiece and a green cord. But for me it has a thousand voices, a thousand faces, a thousand expressions. When it rings, I already know whether it's bringing an ordinary administrative request or is about to announce some new drama, some tragic crime of passion'.]
>
> (*MB* 12–13)

Once again, it is not difficult to find echoes of Futurism. The detective's feelings towards the telephone on his desk can be linked specifically to Filippo Tommaso Marinetti's idea of 'mechanical beauty'.[4] De Vincenzi is a poet-detective because he can feel the poetry of his job, as well as the poetry of humanized mechanical tools. He senses '[la poesia] di quella vecchia stufa, che soffia in tutte le sue giunture, per riscaldar me. E la poesia del telefono!' (*BA* 10) [the poetry of the poor old stove, whose every joint suffers in order to keep me warm. And the poetry of the telephone! (*MB* 13)]. And the chapter ends precisely with the humanized voice of the telephone ringing three times, 'a tre riprese rabbiose, laceranti come tre gridi disperati' (*BA* 13) [Angry and lacerating, like three desperate screams (MB 18)], to announce a new crime somewhere in town.

Despite the preference for a psychological approach to detection, technology and forensic science seemingly play only a marginal role in the resolution of the case. Their presence in the story involves three interconnected aspects: the teasing of both the British model of the private detective and the rationalistic approach to the genre; the agency of non-human actors, such as the pendulum clock, along with that of human characters; and the detective's nuanced relationship with forensic science and the analysis of evidence.

De Angelis builds the character of De Vincenzi in contrast to the idea of a private investigator *à la* Sherlock Holmes. The construction of a new and original Italian detective is also achieved through the negative representation of the British model. When De Angelis states, 'l'uomo ha avuto terribile assillante spasmodico il terrore del crimine elevato a consuetudine di vita quotidiana e subito ha creato un Dio per proteggerlo: il Dio detective, di cui si celebra il culto nei romanzi polizieschi' (*Conferenza* 27) [Man has had a terrible, persistent, spasmodic terror of crime elevated to a habit of daily life, and immediately he created a God to protect him: the detective God, whose worship is celebrated in detective novels (My translation)], he clearly echoes Siegfried Kracauer's theories on detective fiction.[5] According to Kracauer, modern society, dominated solely by formal rationality, created a detective-God as the absolute personification of *ratio*. In the classic detective novel, this ambiguous God-like figure works 'to resolve the mystery of what he himself has not created by the application of formal intellect' (Frisby 11). In *Il banchiere assassinato*, the character of Harrington serves as the perfect example of the British model of the detective. When De Vincenzi meets Count Marchionni for the first time, the nobleman arrives with a private detective as his personal assistant. During the first part of the investigation, De Vincenzi often mocks Harrington by pretending to take his opinions into account. His sarcasm functions to ridicule the British model of the detective-God and the investigative methods based solely on formal rationality. Yet, De Vincenzi does not reject the use of logic, science and evidence in solving a case. What he objects to is the kind of formal logic used by those exceptional investigators of classic detective fiction, who treat reality as a set of equations leading to only one result. Harrington epitomizes this model of investigator with his simplistic conclusions. His self-confidence is irreparably undermined after De Vincenzi discovers that there is a second gun in the apartment, Aurigi has already settled all his debts to the bank, and Count Marchionni knows more than he is letting on. As soon as the private detective is informed of this news, he steps back 'quasi per allontanarsi dal suo cliente' (*BA* 80) [in an effort to separate himself from his client (*MB* 110)]. This model of a private investigator symbolically steps back because it is inconsistent with the Italian approach to detective fiction. Once Harrington moves away from Count Marchionni, he basically leaves the novel for good.

The portrayal of Harrington marks a crucial turning point in the evolution of the genre in Italy, even though he is merely a secondary character. De Angelis' debut novel explicitly rejects the Sherlock Holmes model, aligning with Antonio Gramsci's preference for G.K. Chesterton's Father Brown as a more suitable detective for the Italian *giallo*. In his reflections on detective fiction, Gramsci argues for the superiority of Father Brown over Sherlock Holmes. He believes that Father Brown offers a higher quality of investigation and serves as a better model for Italian writers. Gramsci articulates his thoughts on this popular genre in two letters to Tatiana Schucht in 1930, as well as in sections of his *Quaderni del carcere*, written between 1930 and 1934. According to Gramsci, the Father Brown stories serve as a defence of Catholicism and the Roman clergy, portraying priests as deeply knowledgeable about human nature and emphasizing their function as spiritual guides and intermediaries between humanity and the divine, 'contro lo «scientismo» e la psicologia positivistica del protestante Conan Doyle' (*Letteratura* 138) [against the 'scientism' and positivistic psychology of the Protestant Conan Doyle (*Selections* 371)]. This outspoken stance against Sherlock Holmes and similar investigators, whose approach to detection relies primarily on pure rationality and evidence analysis, eventually influenced the relationship with forensic science and technology in Italian crime fiction, leading to the emergence of what I term *techno-sceptical* sleuths in the following years.

In the character of De Vincenzi, the Italian detective begins to take shape, demonstrating his most distinctive traits, particularly a preference for psychological analysis and a certain scepticism towards the use of scientific methods of investigation. However, in the novel, while the detective almost reluctantly approaches technology, technological tools sometimes seem to seek him out. This becomes evident when the pendulum clock features at a crucial moment in the investigation. Like the telephone and other tools before it, this mechanical object undergoes a process of humanization.

Viewed through the lens of Actor-Network Theory, the pendulum clock in Aurigi's apartment acts as a mediator, facilitating the resolution of the case. According to Latour, 'mediators transform, translate, distort, and modify the meaning or the elements they are supposed to carry' (39). Some non-human actors can shift dynamically from being mere intermediaries to mediators, or vice versa, depending on how they interact with other actors within a network. As Latour argues,

> objects, by the very nature of their connections with humans, quickly shift from being mediators to being intermediaries, counting for one or nothing, no matter how internally complicated they might be. This is why specific tricks have to be invented to *make them talk*, that is, to offer descriptions of themselves, to produce scripts or what they are making others – human or non-humans – do.
>
> (Latour 79)

In De Angelis' novel, the pendulum clock actually speaks to the detective, creating the most important plot twist and reshuffling the roles of the characters in the crime story. Through the narrative humanization of a non-human actor, the novel *makes the clock talk*, transforming it from a simple mechanical tool or piece of furniture into a mediator.

After the initial crime scene analysis, an interrogation of two janitors in the building where the murder occurred marks the true beginning of the investigation. From this moment on, the pendulum clock becomes a crucial element of the story. Although De Vincenzi does not yet know that the clock has been reset, it seems to lure him with an inexplicable force. Once again, a mechanical object is humanized: 'La pendola batteva i minuti, con piccoli scatti sonori. Sembrava un cuore che palpitasse' (*BA* 35) [the clock counted the minutes with audible clicks, like the ticking of a beating heart (*MB* 47)]. The clock also turns into a disquieting presence, drawing De Vincenzi's attention primarily to itself. No longer a simple ornament, the pendulum clock elicits strange emotional reactions in those who observe it: 'A sinistra di quel salotto c'era un caminetto. Sopra il caminetto una pendola. E la pendola batté le ore. Quattro colpi, sonori. De Vincenzi trasalì. Guardò il sestante bianco con quei segni neri e poi Giannetto' (*BA* 40) [There was a fireplace on the left of the room. On the mantelpiece, a pendulum clock. The pendulum marked the hours: four loud strokes. De Vincenzi jumped. He looked at the white clock face with its black points and then at Giannetto (*MB* 54)]. The same magnetic attraction to that mechanical object is reiterated a few pages later, although this time it is another character who observes it: 'Guardò la pendola e trasalì. Segnava le cinque e dieci. Il brigadiere trasse l'orologio di tasca e per qualche minuto rimase a fissare quelle due macchine' (*BA* 42) [He looked at the clock and jumped to his feet. It said ten past five. The sergeant drew his own watch from his pocket and continued looking for several minutes at the two timepieces (*MB* 56)]. Each time De Vincenzi and the other characters gaze at the clock, they experience a palpable sense of discomfort: 'De Vincenzi trasse l'orologio dalla tasca, guardò la pendola e trasalì. Ma non disse nulla e rimise l'orologio in tasca' (*BA* 53) [De Vincenzi took his watch from his pocket and looked at the clock with a jolt. But he said nothing and put the watch back (*MB* 71)]. The same expression ('trasalì', he jumped to his feet) is repeated to convey their reaction to the presence of this object in the room. On a narrative level, the humanization of this mechanical device is nearly complete. In a way, it lacks nothing but words:

> Fu in quell'istante che l'orologio a pendolo, dal caminetto, prese la parola. E batté le ore una dopo l'altra lentamente. De Vincenzi a quel suono sussultò. Fissava l'orologio con occhi accesi, come davanti a una rivelazione. Muoveva le labbra silenziosamente, per contare i colpi. Quasi suggestionati da lui, gli altri seguivano quei suoni e contavano. Anche il conte aveva levata la testa.
>
> (*BA* 84)

> [At that instant, the pendulum clock on the mantelpiece took the floor, slowly striking the hours, one after another. De Vincenzi was startled by the sound. He stared at the clock and his eyes lit up as if in revelation. His lips moved silently, counting the hours. As though he'd suggested it, the others followed the sound and did the same. Even the count had raised his head.]
>
> (*MB* 115)

While the investigation is in full swing and the suspects are gathered in the same room for interrogation, the pendulum clock takes the floor. At this moment, Count Marchionni's daughter, Maria Giovanna, has just confessed to shooting the banker, incriminating herself for his murder. It seems as if the clock speaks directly to De Vincenzi, preventing a tragic mistake. The clock's *voice* awakens the detective's mind, leading him to expose the woman's lies. This technological device proves more effective than any alibi in clearing Giannetto Aurigi, Maria Giovanna and her father from suspicion. It also serves as a Trojan horse to unmask the culprit towards the end of the story, during De Vincenzi's questioning of Aurigi's butler, Giacomo Macchi. During the interrogation, Macchi accidentally reveals that he already knew the clock was off by an hour. In this way, a technological device becomes a mediator by *taking the floor* as a non-human actor, facilitating the flow of information and redefining social roles. The clock's voice also enables the common genre trope in which the detective's questions lead the suspect to contradict themselves during the reconstruction of events. On a narrative level, therefore, a technological device turns into a narrative device, as the agency of this non-human actor facilitates a series of necessary plot twists towards the end of the story, more effectively than any human character has done up to that moment.

The fact that the principal non-human actor in the story is a clock invites further reflection on the agency of technological objects in detective fiction, a genre that has depicted social changes and issues in Italy from its inception to the present. In other words, investigating technology in Italian detective fiction also means exploring the role of technology in the fabric of Italian society.

Kevin Birth describes clocks and time-reckoning tools as 'cognitive artifacts'. Specifically, human-made objects or mechanical devices can be defined as cognitive artefacts when their purpose is to enhance or improve cognition (Hutchins 126). As Birth puts it:

> often cognitive artifacts do not merely substitute for thought, but actually serve to enrich it (….) A feature of some cognitive artifacts is that they contain logics and processes that the users of the artifacts do not understand or even know.
>
> (Birth 7)

These types of objects therefore play a crucial role in shaping cultural models, as they are considered fully reliable based on a shared notion of trustworthiness, whose authority in society derives from the interplay between science, political decisions and economic power. They become integral to a broader concept of society that incorporates non-human elements as well. Time-reckoning tools as cognitive artefacts can also be connected to Mandel's idea of detective fiction as the perfect representation of bourgeois ideology. The genre's early roots are strongly tied not only to the rational and scientific mindset of positivism but also to key principles of capitalist societies in the Western world, such as commodification and reification, which even lead to the idea of reification of death as the very centre of the crime story (Mandel 41). In capitalist societies, the idea of time is often conceptualized as labour time, where time itself becomes a quantifiable commodity. The commodification of time requires that time-measuring instruments be not only reliable but also endowed with a principle of authority, as they are seen as carriers of a truth that is apparently objective and indisputable, because it is socially constructed as such. As Birth remarks, 'the social construction of knowledge, including the ways in which power can shape this construction, can be reflected in the objects by which people think' (8). Clocks as cognitive artefacts fully align with the idea of the social construction of knowledge. The obsession in detective fiction with clocks, calendars, and time-measuring tools reflects bourgeois society's preoccupation with maintaining order. It is no coincidence that the investigation of a crime – especially a murder – often centres around reconstructing the timeline, where time itself becomes a crucial element in solving the mystery. In these narratives, clocks and timepieces are not merely passive objects; they actively shape the plot, reflecting the broader societal obsession with controlling time in both labour and life. As in the real world, it is not necessary for the characters in the text to understand how time-measuring instruments work, how time is calculated or the technology used to do so. What matters is simply that they function as they are supposed to.

In De Angelis' detective novel, the clock *takes the floor* not only to generate a necessary sequence of plot twists, in accordance with the conventions of the genre, leading to the resolution of the mystery. The novel *makes the clock talk* as a non-human actor because it is a pivotal element of the bourgeois and capitalist society in which the criminal event takes place. The death of a banker – embodying 1930s Italian capitalism – disrupts the established order in Milan, the nation's financial hub. Moreover, some members of the Milanese upper bourgeoisie are suspected of the murder. A mechanical time-measuring tool, indispensable to a capitalist society where time is seen as labour time, speaks directly to the detective to catch the criminal and restore the status quo. As a cognitive artefact, the clock represents not only a technological device for measuring time but also a symbol of how capitalist societies manage, control and assign value to human life, labour and, ultimately, death. Whoever moved the hands of the clock did not just tamper with a mechanical instrument but with the order of an entire society, symbolically represented by the violent death of a banker. Therefore, whoever tampered with the clock must be the murderer. And whoever the murderer is, they must necessarily be a marginal figure, integrated into that social context because they are exploitable, a simple and disposable cog in the machine. The butler, Giacomo Macchi, who does not belong to the same social class but serves the Milanese upper bourgeoisie, is therefore the perfect culprit. His time is solely measured as labour time, and tampering with the clock symbolizes a disruption of the status quo.

Beneath what appears to be a clichéd story, in which 'the butler did it', De Angelis crafts a representation of capitalist Italian society during fascism – a regime that, interestingly, is never mentioned throughout the novel, despite De Vincenzi being a police commissioner. The agency of a technological object contributes to solving the mystery. At the same time, as a mediator, it becomes a key element in depicting deeper social dynamics of power that Italian crime fiction analyses and brings to light.

Along with the humanization of mechanical tools and the agency of non-human actors, forensic science and evidence analysis play an important role in unmasking the murderer. When De Vincenzi arrives at Aurigi's apartment, police officers and forensics are already dealing with the crime scene analysis. The second chapter describes their work in collecting the evidence. Although De Vincenzi is no Sherlock Holmes, he is particularly focused on following the correct procedures for analysing the crime scene and detecting potential clues. In this first part of the novel, De Angelis engages with the reader through the description of the crime scene analysis and showcases the poetry of this genre by using technical terms. He also informs the reader about modern methods of investigation, detection and identification procedures, as well as the forensic medical examination of a dead body. This is evident, for instance, in the conversation between De Vincenzi and the Medical Examiner:

> «È morto. La prego, dottore, di voler segnare la posizione esatta del corpo ... Si faccia aiutare da un agente... (...) E la prego, dottore, di spogliarlo e di farmi consegnare gli abiti, procurando che non cada nulla dalle tasche. Ma prima lo esamini bene. Veda se c'è stata lotta ... e da quanto tempo lo hanno ucciso...». (...)
>
> «Approssimativamente, vuol dire. Nessuno può stabilire con esattezza da quanto tempo un uomo è morto. Oppure si potrebbe anche stabilirlo, ma con gli strumenti adatti e prendendo la temperatura dell'ambiente ... e tutte queste cose qui mancano...».
>
> (*BA* 23)

> ['He is dead. I'd ask you, doctor, to make a note of the exact position of the body ... get one of the officers to help you (....) And I'd also ask you, doctor, to undress him – taking care that nothing should fall out of his pockets – and to let me have his clothing. But first, examine him carefully. See if there's been a struggle, and how long ago he was killed.' (...)
>
> 'Approximately, you mean. No one can establish precisely when a man has died. Actually, one could determine it, but only with the right tools and taking the surrounding temperature ... and these things are lacking.']
>
> (*MB* 30–31)

In terms of plausibility, this short exchange is surely lacking. It is unlikely that a coroner would need to be instructed in such detail about the exact procedure for examining a corpse at a crime scene. It is similarly implausible that a highly experienced police officer would be unaware that determining the time of death is nearly impossible without the proper environmental conditions and technical equipment. Most likely, the purpose of this dialogue is to inform the reader about the scientific method of crime scene analysis and the correct forensic procedures. However, in apparent contrast to the precise instructions given to the coroner, De Vincenzi seems indifferent to the almost complete lack of care with which the other officers handle the evidence. When his older colleague, Maccari, notices a small sparkling object on the floor, he picks it up without using a handkerchief or gloves. Maccari ignores the possibility of leaving his fingerprints on a potential piece of evidence, handling and almost playing with the metallic object – a lipstick tube – whose discovery could indicate the presence of another person in the apartment, probably a woman.

Other references to forensic science appear in this preliminary stage of the investigation, further illustrating De Vincenzi's conflicted relationship with modern investigative methods. Before the doctor leaves the crime scene, he updates De Vincenzi on the examination of the victim's body. The medical report is highly technical and specific, particularly when describing the causes of death and in its discussion of ballistics. Furthermore, during a phone call with a colleague, De Vincenzi mentions that the Milan police headquarters are working together to find Aurigi's butler. This phone call is particularly interesting in relation to technology because De Vincenzi refers to the transmission of a 'ritratto parlato' (*BA* 54) via telegraph to all Italian police stations. By using this term, the author alludes to Alphonse Bertillon's biometric research studies on criminal anthropology. Specifically, the Italian expression *ritratto parlato* (identikit) relates to anthropometry, Bertillon's scientific identification system based on physical measurements of the human body. In the late nineteenth century, Bertillon also standardized the use of the mugshot as an investigative tool. Besides fingerprints, thanks to Bertillon, anthropometry and mugshots became the primary scientific systems that police could use to identify a criminal.

It is interesting to note that De Angelis does not provide any further clarification about this procedure. Unlike the detailed description of the medical examination of the dead body, the author here assumes that the reader knows exactly what a *ritratto parlato* is. It is also worth noting that references to forensic science and technological methods of investigation disappear from the story until they resurface in the penultimate chapter. After the initial crime scene analysis, the reader receives no further information about the analysis of evidence, ballistics or the medical report, as the investigation focuses solely on the detective's psychological analysis of people's behaviour.

Although De Vincenzi is more interested in exploring 'the mystery of the human soul' (*MB* 13) rather than chasing down material evidence, the investigation reaches a standstill. Despite the value of the psychoanalytical approach, it seems that shifting back to the analysis of evidence is the only possible way to move the story forward. After accidentally finding part of the material evidence in his coat pockets, De Vincenzi uses a stratagem to obtain the butler's fingerprints for comparison with those found on the gun. After many pages dedicated to exploring the characters' psychological anxieties and uncovering their skeletons in the closet, the analysis of evidence is now presented as the most effective means to achieve the detective's goal. Unlike his earlier indifference to preserving evidence, De Vincenzi now focuses solely on forensics to find the suspect. Psychological aspects give way to scientific proof. Whereas De Vincenzi initially overlooked Maccari handling objects without gloves and contaminating potential evidence, he now takes all necessary precautions to pull prints, aiming to provide irrefutable proof of Giacomo Macchi's guilt.

> «Prendete quella rivoltella … ma state ben attento di non togliere il fazzoletto e di non toccarla…» (…) toltosi un altro fazzoletto dal taschino della giacca, ne ravvolse il bicchiere e lo porse all'agente: «E prendete questo…». La voce gli si era fatta dura: «Tanto sulla rivoltella quanto sul bicchiere vi sono impronte digitali. Andate subito al Gabinetto della Polizia Scientifica e fatemele rilevare … Ma presto! Tra un'ora voglio le fotografie».
>
> (*BA* 122)

> ['Take the revolver – but be very careful not to remove the handkerchief and not to touch it.' (…) He drew another handkerchief from the small pocket of his jacket, wrapped the glass in it and held it out to the officer. 'Take this.' His voice had become hard. 'There are fingerprints on the revolver as well as the glass. Go to the Forensics Office right now and have them lifted. Quickly! I want the photographs within an hour.']
>
> (*MB* 167–8).

Technology and forensic science thus play a pivotal role in De Angelis' detective novel, even though the author suggests a preference for the psychological analysis of characters. In a way, the narrative structure unfolds as a psychological drama framed within a classic whodunit. The investigation begins with the crime scene analysis and only picks up again towards the end after De Vincenzi accidentally finds part of the evidence in his coat pocket. Essentially, nothing significant happens in the central part of the novel, aside from the active role of the pendulum clock as a non-human actor.

De Angelis' *Il banchiere assassinato* is not only an excellent detective story, but it also lays the foundation for a distinctly Italian approach to crime fiction. The novel translates the author's theories on the genre into an original Italian form. Furthermore, it aligns with Gramsci's critical reflections on detective fiction, particularly his assertion of the superiority of G.K. Chesterton's Father Brown over the analytical and hyper-rational model of investigator epitomized by Arthur Conan Doyle's Sherlock Holmes. De Angelis could not have been aware of Gramsci's ideas when writing both his debut novel and the *Conferenza sul giallo*. However, it is evident that he directly engages with Chesterton's short essay on detective fiction, first published in 1902, particularly the claim that detective fiction 'is the earliest and only form of popular literature in which is expressed some sense of the poetry of modern life' (Chesterton 75). Fascinated by the cultural trends of the time, De Angelis combines in his novel the psychological introspection and the Futurist fascination with machinery, creating a poet-detective who is primarily concerned with delving into 'the mystery of the human soul'. However, despite De Vincenzi's ambivalent relationship with modern methods of investigation, technology and forensic science play an important role in the story. In particular, through a process of humanization, the pendulum clock transforms into a fundamental narrative device for solving the mystery. More importantly, as a non-human mediator, it facilitates the redefinition of social roles within the story, shedding light on deeper social dynamics in Italy at the time. This use of technology would later become a hallmark for many politically engaged Italian authors in crime fiction as a form of social commentary.

Notes

1 De Angelis' crime stories were reevaluated only many years after his death thanks to Oreste De Buono, who worked on a new edition of three selected novels in the volume titled *Il commissario De Vincenzi*, published by Feltrinelli in 1963. Several more detective stories were republished by Garzanti and Sonzogno in the 1970s, especially due to the popularity of the television series *Il commissario De Vincenzi* (1974–1977), starring actor Paolo Stoppa as the detective. More recently, some of the novels were republished by Sellerio in the early 2000s and by Mondadori in the collection 'Il Giallo Mondadori: Classici Oro'.

2 Henceforth *Conferenza* in in-text citations.

3 Henceforth, *BA* (and *MB* for the English version) in in-text citations.

4 See especially *L'uomo moltiplicato e il Regno della Macchina*, where Marinetti exalts the idea of mechanical beauty by 'promoting love of the machine' (*Critical Writings* 85) as an affectionate feeling and act of tenderness similar to that of a man lovingly caressing his beloved (*Filippo Tommaso Marinetti* 39).

5 Siegfried Kracauer's essay *Der Detektiv-Roman* was written between 1922 and 1927, but it was published posthumously only in 1971. I refer to the 2011 Italian edition, titled *Il romanzo poliziesco*, translated by Renato Cristin.

References

Birth, Kevin K. *Objects of Time: How Things Shape Temporality*. Palgrave Macmillan, 2012.

Brunetti, Bruno. *Augusto De Angelis. Uno studio in giallo*. Edizioni Graphis, 1994.

Carloni, Massimo. *L'Italia in giallo: Geografia e storia del giallo italiano contemporaneo*. Edizioni Diabasis, 1994.

Centini, Massimo. *Storia della criminologia e dei metodi investigativi*. Diarkos, 2022.

Chesterton, G. K. (Gilbert Keith). *The Defendant*. Edited by Dale Ahlquist, Dover Publications, 2012.

Cremante, Renzo, editor. *Le figure del delitto. Il libro poliziesco in Italia dalle origini a oggi*. Grafis, 1989.

Crovi, Luca. *Storia del giallo italiano*. Marsilio, 2020.

Crovi, Luca. *Tutti i colori del giallo. Il giallo italiano da De Marchi a Scerbanenco a Camilleri*. Marsilio, 2002.

De Angelis, Augusto. *Il banchiere assassinato*. Elliot, 2018.

De Angelis, Augusto. "Conferenza sul giallo (in tempi neri)." *La Lettura*, vol. 47, Mar. 1980, pp. 27–44.

De Angelis, Augusto. *The Murdered Banker*. Translated by Jill Foulston, Pushkin Vertigo, 2016.

Del Buono, Oreste. "Introduzione." *Il Commissario De Vincenzi*, edited by Augusto De Angelis, Feltrinelli, 1974, pp. 9–15.

Dunnett, Jane. "The Emergence of a New Literary Genre in Interwar Italy." *Italian Crime Fiction*, edited by Giuliana Pieri, U Wales P, 2011, pp. 6–26.

Frisby, David. "Between the Spheres: Siegfried Kracauer and the Detective Novel." *Theory, Culture & Society*, vol. 9, no. 2, 1992, pp. 1–22.

Gramsci, Antonio. *Letteratura e vita nazionale*. Editori Riuniti, 1991.

Gramsci, Antonio. *Selections from Cultural Writings*. Translated by William Boelhower and Edited by David Forgacs and Geoffrey Nowell-Smith, Harvard UP, 1985.

Hutchins, Edwin. "Cognitive Artifacts." *MIT Encyclopedia of the Cognitive Sciences*, edited by Robert A. Wilson and Frank C. Keil, MIT Press, 1999, pp. 126–7.

Kracauer, Siegfried. *Il romanzo poliziesco*. Translated by Renato Cristin, SE, 2011.

Latour, Bruno. *Reassembling the Social: An Introduction to Actor-Network Theory*. Oxford UP, 2005.

Mandel, Ernest. *Delightful Murder: A Social History of the Crime Story*. Pluto Press, 1984.

Marinetti, Filippo Tommaso. *Critical Writings*. Translated by Doug Thompson and Edited by Gunter Berghaus, Farrar, Straus and Giroux, 2006.

Marinetti, Filippo Tommaso. *Filippo Tommaso Marinetti e il futurismo*. Edited by Luciano De Maria, Mondadori, 2000.

Petronio, Giuseppe. *Il punto sul romanzo poliziesco*. Laterza, 1985.

Pezzotti, Barbara. *Politics and Society in Italian Crime Fiction*. McFarland, 2014.

Pistelli, Maurizio. *Un secolo in giallo. Storia del poliziesco italiano (1860–1960)*. Donzelli, 2006.

Rambelli, Loris. "Acculturazione di un genere letterario: Il detective, l'analista Italiano." *Lingua e Stile*, vol. 10, 1975, pp. 97–124.

Rambelli, Loris. *Storia del "giallo" italiano*. Garzanti, 1979.

Sangiorgi, Marco. "Il fascismo e il giallo italiano." *Il giallo italiano come nuovo romanzo sociale*, edited by Marco Sangiorgi and Luca Telò, Longo, 2004, pp. 117–52.

Somigli, Luca. "The Realism of Detective Fiction: Augusto De Angelis, Theorist of the Italian Giallo." *Symposium*, vol. 59, no. 2, Summer 2005, pp. 70–89.

Todorov, Zvetan. "The Typology of Detective Fiction." *The Poetics of Prose*. Cornell UP, 1977, pp. 42–52.

3 On the Threshold of Life and Death

Guido Cavalcanti and the Medieval Automaton

Rebecca Reilly

Introduction

It is often assumed that reflection on artificial life and the relation of the automaton to the human is a purely modern concern, connected to technological developments from the Industrial Revolution to today's digital age. In fact, such reflections extend throughout history. In classical antiquity, Aristotle invoked 'automatic puppets' to explain the functioning of animal movement. Aristotle explains that the motion of 'automatic puppets', though similar to animal motion, lacks the qualitative change that occurs in animals through perception, sensation and imagination (42–5). In the seventeenth century, René Descartes asserted a more extensive similarity between animals and automata, claiming that humans are the only animals truly different from machines by virtue of their rational ensoulment (99–108). Despite the centuries separating them, for both thinkers (and for many others), the automaton functions as a conceptual tool, bringing into relief the qualities that define human life, above all rationality. Nonetheless, drawing a parallel between human and machine threatens to blur the boundaries between them, to demolish this sense of human exceptionality, to confound distinctions between natural/artificial and sentient/non-sentient.

This chapter examines the use of the automaton as a tool for defining (and blurring) the boundaries of human life in a text from the period between Aristotle and Descartes. This is not an explicitly philosophical text but a poetic one, a sonnet by the late medieval Florentine poet Guido Cavalcanti (c. 1259–1300). Cavalcanti is best known for his appearances in the works of Dante (c. 1265–1321) and Boccaccio (1313–75). Yet his lyric poems, of which around 50 survive, are strikingly original in the Italian lyric panorama and had significant influence on subsequent poetic production. Cavalcanti's originality is often detected in his precociously 'modern' lyric subjectivity. As Roberto Antonelli comments: 'it is with him that modern lyric [...] begins, understood as a self-conscious, poetic investigation of the inner self' (3).[1] Another major innovation in the

DOI: 10.4324/9781003545910-4

European vernacular lyric tradition embodied not only by Cavalcanti but, more broadly, by the lyric production of the circle of Florentine poets of which he, along with his younger colleague Dante, was a part, relates to philosophical content. As Gregory Stone notes, these poets were 'writing love lyric that is not just occasionally adorned with but essentially *is* philosophical discourse'; the lyric form thus becomes a vehicle for reflection on the nature of human life (*Guido* 4).

It is perhaps within this philosophical-poetic project that we should situate Cavalcanti's evocation in one of his sonnets (VIII, pp. 72–5) of the figure of an automaton as a metaphorical depiction of the lyric subject's suffering in love:[2]

Tu m'hai sì piena di dolor la mente
che l'anima si briga di partire
e li sospir' che manda 'l cor dolente
mostrano agli occhi che non può soffrire

Amor, che lo tuo grande valor sente,
dice: 'E' mi duol che ti convien morire
per questa bella donna, ché nïente
par che Pietate di te voglia udire'.

I' vo come colui ch'è fuor di vita,
che pare, a chi lo sguarda, ch'omo sia
fatto di rame o di pietra o di legno

che si conduca sol per maestria
e porti ne lo core una ferita
che sia, com'egli è morto, aperto segno.

You have filled my mind with such pain
that my soul prepares to take its leave
and the sighs that my pained heart sends out
show visibly that it can suffer no longer.

Love, who feels your great power,
says: 'It pains me that you must die,
for this beautiful lady, who seems
to want to hear nothing of pity for you'.

I go about like one who is empty of life,
Who seems, to whoever looks upon him, to be a man
Made of copper, or stone, or wood

Who moves only through artifice
And who bears in his heart a wound
That is an open sign of his death.

This sonnet presents a striking scene of living death: the flight of the lyric subject's 'anima' (*soul*, v. 2) from his body reduces him to a walking corpse. While the verbs 'vo' (*I go*, v. 9) and 'si conduca' (*walks*, v. 12) indicate mechanical motion, his body is emptied of inner life, explicitly 'morto' (*dead*, v. 14), 'fuor di vita' (*lifeless*, v. 9). Cavalcanti's references to inanimate materials ('rame' [*copper*], 'pietra' [*stone*] and 'legno' [*wood*], v. 11) emphasize the lifelessness of this body, while the caesura in the line 'pare […] ch'omo sia' (*it seems […] to be a man*, v. 10) indicates that this body only seems to be, but is not quite, human. That this figure resembles an automaton is evident from the reference to craft materials and from the term 'maestria' (v. 12), where 'maestria' connotes, in Domenico de Robertis' words, 'artifice (of the master who fabricated it), contraption, mechanism' (in Cavalcanti, *Rime* 29).[3]

Commentators have consistently identified the final two tercets of this sonnet as evoking an automaton.[4] However, excluding a brief essay by Paola Vecchi Galli, there has been little scholarship contextualizing Cavalcanti's automaton within ancient and medieval technological traditions or within the thirteenth-century Italian lyric tradition. This chapter provides an in-depth contextual reading of Cavalcanti's sonnet. I begin by expanding on Vecchi Galli's observations in her essay, first surveying the panorama of medieval European understandings of automata that may have been available to Cavalcanti and, second, outlining parallels between Cavalcanti's imagery and that of earlier lyric poets. I then situate this imagery within the physiology of love as an irrational, lethal passion that undergirds Cavalcanti's lyric oeuvre. Finally, I turn to a neglected facet of Cavalcantian poetic imagery: its religious symbolism. The automaton of sonnet VIII draws not so much on a modern 'scientific' understanding of artificial life but recalls various automaton-like objects from contemporary devotional culture, situating the lyric subject in a liminally mortal state that parodically recalls Christ. By comparing his lyric subject simultaneously to an automaton and to Christ, Cavalcanti's sonnet offers an ambivalent exploration of medieval Christian ontologies of matter and embodiment and their vacillation between heavenly/eternal and earthly/organic definitions of life.

Automata between Life and Literature in the Middle Ages

Though artificial life has been a philosophical and literary concern from at least antiquity to the present day, there has often been a discrepancy between the automata envisioned in art and literature and the technological mechanisms for building automata available in any given period. Adrienne Mayor's survey of automata in ancient Greek myth establishes that, long before technological advancements in Hellenistic Alexandria made the production of automata possible, such objects were imaginable, just as modern science fiction imagines objects beyond current technological capacities (1, 96). The notion that literature's imaginative scope regarding automata exceeds what is technologically possible applies to the medieval West, where the proliferation of automata in literature far outstripped the available technical knowledge for producing similar objects.

The medieval technological tradition related to automata can be traced back to the ancient world, specifically to Hellenistic Alexandria. In this period, engineers and inventors, particularly Philo of Byzantium (c. 280–220 BCE) and Hero of Alexandria (c. 10–70 CE), produced designs for functioning automata, machines imitating human and animal life, using hydraulic and mechanical techniques (Brumbaugh; Truitt, *Medieval* 3–5; Mayor 199–201). The work of these Alexandrian writers was, as Elly Truitt notes, preserved in the Byzantine world, and began to be absorbed and expanded upon in the Islamic world from the eighth century (*Medieval* 19-26). The earliest Western description of an automaton relates to an object which hailed from this Arabic tradition: in 807, a water clock embellished with mechanized horsemen was reportedly gifted to Charlemagne by the Caliph of Baghdad (Truitt, *Medieval* 20–1). The existence of automata was therefore known to the Western world, yet they were not constructed in the West until the thirteenth century. According to Truitt, the first automata constructed in the West were the robotic animals built in the garden of the Chateau of Hesdin around 1300, commissioned by Count Robert II of Artois (*Medieval* 117, 123). Prior to this, Villard de Honnecourt (active 1225–50) recorded rudimentary designs for automata in his notebooks, yet it seems unlikely that they were ever actually built (Truitt, *Medieval* 118–22). Joel Pattison notes that it is possible that automata were constructed shortly prior to this in Genoa and/or Sicily (4–5); additionally, as will be discussed later in this chapter, one extant automaton was constructed much earlier in the century in Christian Spain. Yet the building of automata in the Western world did not take place before the thirteenth century, lagging centuries behind Arabic and Byzantine mechanics. Treatises on the engineering of automata – those of the Hellenistic world and those of Arabic and Byzantine successors – remained, until this late period, unavailable to the Latin West.

Nonetheless, the presence of automata throughout Western literary culture in this period remains extensive. Travel reports of the journeys of European explorers and diplomats to Islamic and Byzantine courts tell of marvellous automata found there, constructed from expensive materials and with technological expertise unknown to the West. Fictional romances and *chansons de geste* similarly incorporate automata into their narratives, often associating them with foreign lands or the ancient world (Daston and Park 90–4; Truitt, *Medieval* 12–39, 63–8). Such depictions of automata rely on a medieval 'topography of wonder', to use the terminology of Lorraine Daston and Katharine Park, whereby examples of the marvellous were imagined to be concentrated at the margins of the Christian world (21–66). Lacking the scientific knowledge required to properly analyse the mechanical function of these machines, medieval European writers explain automata as the fruit of arcane natural philosophical and/or magical knowledge of dubious moral legitimacy due to their non-Christian provenance (Truitt, 'Trei poëte'; *Medieval* 40–68). Truitt notes that the symbolic status of automata as 'liminal' objects with respect to the Christian world means that they are often utilized within literary representations to 'comment on categories, often calling them into question by their very existence' through the ontological challenge that their existence poses to boundaries between natural and artificial, life and death, licit and illicit knowledge (*Medieval* 9). However, in present critical discussions of the function of automata as symbols of knowledge, otherness and liminality in medieval European literature, among which Truitt's work stands out, the focus has been on French language materials and on narrative works, particularly romances and *chansons de geste*: *Le Voyage de Charlemagne* (c. 1140), *Le Conte de Floire et Blancheflor* (c. 1150–1170), *Le Roman de Troie* (c. 1165) and *Le Roman d'Eneas* (c. 1160), to name a few. It is worth considering how Cavalcanti's situation in an Italian rather than French cultural context and use of lyric rather than narrative poetic forms may inflect his depiction of automata, distinguishing him from the poets discussed by Truitt.

The Tuscan poets of the late twelfth century, including Cavalcanti, were inheritors of the Italian vernacular poetic practices of the 'Sicilian School', active in the first half of the century in Palermo at the Court of Frederick II, whose practices drew on the legacy of Occitan troubadours (Abulafia 270–9; Barolini 23–45). Sicily under Frederick II was a centre of poetic development in the Italian vernacular but also an important site for the transmission of Arabic and Byzantine science to the Latin West. Maria Luisa Ardizzone suggests that these scientific-philosophical traditions formed an important background and influence for Sicilian School poets and argues that Cavalcanti inherited similar Arabic philosophical tendencies, as evidenced by Cavalcanti's extensive interest in the re-interpretations of Aristotle by the twelfth-century Islamic Andalusian thinker Averröes (*Guido* 13–14). Frederick II's Sicilian court was a site for the transmission of Arabic science not only through texts but also objects, specifically automata. Sources suggest that Frederick received automata as gifts from Muslim rulers: an astrological water clock and a metal tree with artificial songbirds (Abulafia 266–7; Farmer 115; Truitt, *Medieval* 125–6). Pattison notes that while some sources claim that Frederick received the tree automaton from the sultan of Egypt, others claim that Frederick sent the tree to the sultan, raising the possibility that automata were constructed in Sicily decades before their construction elsewhere in Western Europe (Pattison 5, n. 16). Research on the presence and construction of automata in Frederick II's Sicily remains cursory, and an in-depth investigation of this topic exceeds the scope of this chapter. However, the importance of Sicily as a site of cultural exchange, particularly regarding the transmission of scientific knowledge from Arabic (and Byzantine) lands to the West, raises the possibility that Cavalcanti's practical knowledge of automata, as potentially constructible rather than as esoteric literary objects, may have been more advanced than that of twelfth-century French narrative poets.

Nevertheless, the extent of Cavalcanti's scientific knowledge regarding the construction of automata is impossible to decipher from sonnet VIII. Cavalcanti does not explain the functioning of automata in any way, neither speculating on their origins in arcane, quasi-magical natural philosophical knowledge (as occurs in the romances discussed by Truitt) nor correcting this with practical, mechanistic understandings. Compared to the extensive descriptions of automata in romances, *chansons de geste* and travel narratives, in Cavalcanti's sonnet, the image is reduced to its bare bones. This verbal economy is related to the exigencies of lyric, which, particularly in short forms like the sonnet, does not permit the extended, reflective description of narrative. For Jonathan Culler, the effectiveness of lyric discourse 'depends on its success in making words memorable' (130). Thus, in Cavalcanti's sonnet, the image of the automaton is reduced to the elements that most effectively produce a striking, memorable image in the reader's mind.

The memorability of Cavalcanti's sonnet is perhaps achieved by harnessing the 'uncanny' sentiments aroused by human-like automata, as theorized by modern robotics. Masahiro Mori posited that we initially feel more at ease with automata as they become more lifelike, but at a certain point, this lifelikeness evokes horror. This sudden dip in comfort level is what Mori famously referred to as the 'uncanny valley'. For Mori, this sense of the uncanny is evoked by nearly-but-not-quite lifelike robots, puppets and prosthetics, but also by entities like zombies that are similarly liminal in their resemblance to living humans. Recent robotics researchers have subtly altered Mori's theory with evidence from controlled testing. A study by Karl MacDorman and Hiroshi Ishiguro found that, while Mori theorized that comfort with robots increases as they become more human-like before dropping into the 'uncanny valley', in practice, discomfort with robots rises as this likeness increases. The 'uncanny valley' is thus not a sudden dip but the culmination of a gradually increasing unease. As Minsoo Kang describes, MacDorman and Ishiguro thus observe not so much an 'uncanny valley' but an 'eerie peak' with its climax 'in that middle range when there is a maximum level of uncertainty about the nature of the thing observed' (Kang 50). As Kang explains, the 'uncanny' or 'eerie' sensation provoked by such liminal automata is explainable by the fact that they are registered as 'dead things doing things they should not be doing': 'the uncanny automaton is more akin to a zombie, a living mummy, or a vampire, as it crosses the categories of animate/inanimate, living/dead [...] and so threatens our reality schema in a particularly frightening way' (52). For Kang, it is this life-death liminality embodied by the automaton that makes it an object of both fascination and horror. These strong affective reactions to the uncanny, liminal automaton are channelled in Cavalcanti's sonnet into evoking for readers a visceral and memorable image of a subject on the boundaries between life and death in the extremity of his suffering in love.

The use of the automaton in medieval literature as a symbol of life-death liminality is outlined by Truitt. Truitt notes that, in medieval romances, 'automata are found at liminal spaces [...] their functions are surveillance and discipline, which signal not only the liminal status of automata themselves, but also the ways in which they enforce boundaries of epistemological legitimacy' ('Trei poëte' 172). This is particularly visible when automata appear as guardians of tombs, upholding boundaries between living and dead but also indicating the precariousness of these boundaries (Truitt, 'Fictions'; Truitt, *Medieval* 96–115). In one romance that Truitt cites, the *Roman d'Eneas*, the tomb of Camille is guarded by an automaton-archer, ready to shoot if the tomb is disturbed. The automaton patrols the separation of living from dead, but also confounds life-death boundaries by protecting Camille's uncorrupted, embalmed body from the entrance of air into the tomb and thus from decay ('Fictions' 2; *Medieval* 102–4). In another romance, *Le Conte de Floire et Blancheflor*, Floire, a Saracen prince, falls in love with Blancheflor, a Christian prisoner. Floire's disapproving father sells Blancheflor to the emir of Babylon but tells his son that she has died. To uphold this fiction, the king builds a mausoleum for Blancheflor, decorated with two moving, speaking marble statues that share an eerie likeness to the real Floire and Blancheflor. As Truitt argues, these automata trouble the boundaries between life and death in both directions: 'the lifelike, animated statues recall the dead as they were when alive, illustrating a fiction of life, even as the tomb, of a person who is not actually dead, offers the fiction of death' ('Fictions' 3).

In evoking the liminally mortal automaton, Cavalcanti, similarly to the romances discussed by Truitt, explores the precariousness of boundaries between life and death. Sonnet VIII repeatedly underlines the death-like state suffered by the lyric subject, accentuating this through the parallel with the automaton. While in the romances discussed by Truitt, the automaton remains ontologically distinct from 'human' characters in the narrative, in Cavalcanti's lyric, the automaton becomes a metaphorical figuration of the human subject itself, conveying the lyric 'I's reduction to a state of inner death by the painful experience of love. The evocation of the automaton in sonnet VIII seeks greater affective immediacy than the extended descriptions of romance narratives: Cavalcanti harnesses the automaton's 'uncanny' qualities to provoke an instantaneous response of horror and awe from readers, thereby performatively expressing the extremity of the lyric subject's suffering in love.

Love as Death in the Italian Lyric Tradition and in Cavalcanti's Oeuvre

Scholars including Paolo Rigo and Isabelle Battesti highlight the pervasiveness of the motif of love as 'living death' throughout medieval European vernacular love lyric. This can be traced back to the Occitan troubadours. In a *canso* (XXII) by the troubadour Folquet de Marseille (c. 1150–1231), we read:

> Donc muer e viu? Non, mas mos cors cocios
> mor e revieu de cosir amoros (vv. 7–8, p. 95).[5]
>
> *So do I live and die at once? No, but my yearning heart*
> *both dies and is brought back to life by these amorous thoughts.*

Folquet's poem was translated/adapted into Italian by Giacomo da Lentini (c. 1210–60), pre-eminent poet of the 'Sicilian School'. In Giacomo's version, 'Madonna, dir vo voglio', (*Lady, I wish to tell you,* I), Folquet's emphasis on love as death (or near-death) is replicated and accentuated, as Giacomo closely follows the troubadour's original lines while placing even greater emphasis on a lexicon of death:

> Adunque morire' eo?
> No, ma lo core meo
> more spesso e più forte
> che no faria di morte– naturale (vv. 9–12, p. 3).[6]
>
> *So am I dying?*
> *No, but my heart*
> *dies more often and more powerfull*
> *than it would if it were to die naturally.*

The depiction of love as fatal would remain decisive when Italian lyric shifted North, from Sicily to Tuscany, in the second half of the thirteenth century. The automaton metaphor was introduced into the Italian lyric tradition at this point as a variant on this motif of love as living death. The first appearance of such imagery occurs in a sonnet (IV) by the Bolognese poet Guido Guinizzelli (c. 1225–76), a poet of the generation prior to Cavalcanti:

Remango come statüa d'ottono
ove vita né spirito non ricorre,
se non che la figura d'omo rende (vv. 12–4, pp. 42–3).[7]

I am left like a brass statue,
to which neither life nor spirit return
yet that depicts the figure of a man.

In Guinizzelli's sonnet, the lover is a statue rather than an automaton proper, and, according to Vecchi Galli, the fact that, in Cavalcanti's sonnet, the lover is not merely a statue but a *moving* statue distinguishes his image from Guinizzelli's precedent and renders it 'a true hapax in poetry written in the time of Dante' (23).[8]

Though unmentioned by Vecchi Galli, another poet prior to Cavalcanti did develop this imagery into something more closely resembling an automaton. In a sonnet (CXXXIV) by Guittone d'Arezzo (c. 1235–94), the lady's departure causes such suffering that the lover is reduced to a death-like state resembling a wooden puppet:

Ed ha lassato il corpo quasi morto
che va e vene, ma non po parlare
ed ogn'om guarda, né vede chi sia

Ma par che viva come legno torto (vv. 9–12, p. 206).[9]

And [Love] has left my body almost dead,
so that it comes and goes, but cannot speak,
and everyone gazes upon it but cannot see who it is

But it seems to live as if it were a piece of twisted wood.

Like Cavalcanti, who invokes craft materials to describe his lyric subject, Guittone compares the body of his suffering lover-subject to 'legno' (*wood*, v. 12). Just as Cavalcanti's automaton-lover continues to move superficially, Guittone's lyric 'I' performs some mechanical functions of human embodiment: 'va e vene' (*he comes and goes*, v. 10). Moreover, both poets explicitly connect the automaton metaphor to the death-like state of the lyric subject suffering in love.

Rigo (118–21) gestures towards an explanation for the rationale behind the lyric motif of 'living death', particularly as it appears in Cavalcanti's sonnet, found in Book IV of Dante's *Convivio,* as a gloss to the line of his canzone 'Le dolci rimi d'amor' (The sweet rhymes of love, v.1, p. 253): 'e tocca a tal, ch'è morto e va per terra' (*of him it can be said that he is dead and walks on earth*, v. 40, p. 255).[10] Dante explains, based on the Aristotelian distinction between vegetative, animal and rational life, that, for humans, 'vivere [...] è ragione usare' (*to live is to use one's reason,* IV .vii.11, p. 305). For Dante, humans that renounce rational thought are not properly human: 'rispondo che è morto [uomo] e rimaso bestia' (*to renounce the use of reason is to renounce being, and thus is the same as being dead*, IV.vii.14, p. 305). As Rigo signals, the 'living death' invoked by lyric love poets is a similar death of reason: love is depicted as extinguishing the cognitive activity that defines the human, reducing the subject to a purely corporeal life of physical passion. Indeed, the disruption of rational thought and faculties of judgement was central to medico-philosophical definitions of lovesickness on which lyric poets often drew (Ciavolella 97–101). The next lines of Dante's canzone invoke the figure of an automaton as a means of distinguishing rational human life from mere 'animation': 'Chi diffinisce: 'Omo è legno animato'/ [...] dice non vero' (*Whoever defines man as animated wood/ says something untrue*, vv. 41–2, p. 255). In his prose commentary, Dante explains that those who claim that humans are merely animated wood are incorrect 'in quanto dice "animato", non dicendo "razionale", che è differenza per la quale [l']uomo dalla bestia si parte' (*insofar as they say 'animated' instead of 'rational', as the latter is the difference that separates human from beast,* IV.x.4, p. 323). It is thus reason that sets humans apart not only from animals but also automata, whose 'animation' lacks the rationality that ideally governs human action.

The notion that the 'death' invoked by lyric love poetry is a death of reason, an intellectual or spiritual rather than a bodily death, resonates with Cavalcanti's adherence to the natural philosophical principles of the Arabic thinker Averröes. Cavalcanti's Averroism was posited by Bruno Nardi, and scholars since, including Maria Corti (3–37), Ardizzone (*Guido*) and Stone ('Animals'; *Guido* 165–258), have offered readings of the poet's lyrics, especially his *canzone* 'Donna me prega' (A lady bids me XXVII) confirming this philosophical stance. As Ardizzone explains, theories of Cavalcanti's Averroism are rooted in the fact that he apparently posits, in 'Donna me prega', that humans are excluded from intellection, which occurs in a supra-individual Intellect (*Guido* 103–33). Because humans do not partake in intellection, for Averroists, human experience is purely sensory and bodily, on a continuum with animal life.

This was controversial from a Christian perspective, implying a lack of belief in the immortality of the human soul that led to ecclesiastical condemnations (Ardizzone, *Guido* 71–102). Cavalcanti's apparent adherence to such Averroist perspectives thus helps to contextualize Dante's allusion to Cavalcanti in the context of his depiction of the sin of heresy in canto X (vv. 52–72) of *Inferno*. Scholars have tended to argue that the exclusion of humans from intellection in Averroist perspectives means that, for Cavalcanti, love is irrational. Stone posits, conversely, that Cavalcanti understands love as potentially rational (although not intellective), since reason is, for Averroists, the highest power of the human soul, but is understood as a variant on imagination rather than on intellection, and thus pertains to the sensitive faculty of the soul and to the material sphere (Stone, 'Animals' 1280). Through this rational level of the imagination which enables the generation of 'phantasia' (imaginary forms of materially perceived objects), humans may participate indirectly in intellection by offering sensory data to be abstracted by the supra-human Intellect (Stone, 'Animals' 1272).

Nonetheless, even if, as Stone posits, it is the case that love *can* pertain to reason, for Cavalcanti, rational love is uncommon. In 'Donna me prega', love is described as follows:

> Di sua potenza segue spesso morte
> Se forte — la vertù fosse impedita,
> la quale aìta — la contraria via:
> non perché oppost' a naturale sia;
> ma quanto che da buon perfetto tort'è
> per sorte — non pò dire om ch'aggia vita,
> che stabilita — non ha segnoria (vv. 35–41, pp. 157–8).
>
> *Its power often leads to death,*
> *if rational virtue were to be strongly impeded,*
> *since this virtue would counsel alternative paths.*
> *This is not because love is unnatural,*
> *but because it turns one away from the perfect good*
> *by fate— one cannot be said to be alive*
> *when love is established— one has no self-control.*

Love becomes deadly if 'la vertù fosse impedita' *([rational] virtue were impeded,* v. 37), that is, if an excess of passion prevents the generation of 'phantasia' to be abstracted by the supra-human Intellect. The death often resulting from love is described as a turn away from the 'buon perfetto' (*perfect good,* v. 39), from rational imagination, the highest part of the sensitive soul and thus of human life (Ardizzone, *Guido* 127–8). Moreover, the lifeless lover is described as lacking rational control over his own being ('non ha segnoria', *has no self-control,* v. 41), and, for this reason, *cannot be said to be alive* (v. 40). Giorgio Inglese thus describes the death that results from love, according to the *canzone,* as a '"spiritual death", that is, interruption of that intellectual activity that is the only true *life* of the human' (in Cavalcanti, *Rime* 148).[11] However, since, according to Averroists, human imagination and reason are part of the sensitive, corporeal level of being (while intellection resides beyond the human), this 'spiritual' death is intertwined with physical disruption. As Ardizzone outlines, the excess of passion resulting from love is frequently depicted in Cavalcanti's oeuvre as effecting bodily disharmony which, in turn, blocks the subject's cognitive powers: 'the destruction of the body is also related to the death of the imagination, and [...] the real death is that of the imagination' (*Guido* 34). Or, as Isabelle Battesti writes, 'in Guido's lyric corpus, this ontological non-life is never clearly distinguished from physical death' (118–9).[12]

The death of the lyric subject is enacted in Cavalcanti's oeuvre as the reduction of the body to an empty vessel outside of the subject's control, lacking individuality or interiority, even if it uncannily continues to move mechanically. While in sonnet VIII, as the first quatrain describes, it is the 'anima' (*soul*) that leaves the poet's body, elsewhere in Cavalcanti's oeuvre, this is replaced with the metonym of 'spiriti' (*spirits*). As Heather Webb explains, according to medieval physiology, 'spiriti' were an 'intermediary [...] between body and soul' (30), so were vital for human life in both its physical and intellectual-spiritual dimensions. The dramatic scenario of the flight of 'spiriti' from the body through a 'wound' produced by the subject's perception of the beloved lady is, as Webb notes, omnipresent throughout Cavalcanti's lyric poetry (74–8).[13] This enacts a death that is simultaneously physical and spiritual: 'spiriti' were ideally supposed to mediate and maintain the bond between body and soul necessary for human life, so their flight signifies the dissolution of this bond and the separation of body from soul. Indeed, as Nancy Caciola notes, the separation of soul from body was the primary late medieval definition of death (72). The flight of the 'spiriti' is not only explicitly narrated in Cavalcanti's lyrics but also enacted through body language, particularly the sigh, an exhalation of 'spiriti' in breath that, according to Arabic medical discourse, may cool a heart overheated from emotion (Ardizzone, *Guido* 22–4). This is prominent in sonnet VI (pp. 64–7), where the apostrophe 'deh' is repeated four times (vv. 1, 5, 7, 12), enacting the physical event of the sigh through the discharge of the breath required to pronounce 'deh'. If this exhalation of 'spiriti' is necessary to soothe the overheating heart, it also, if taken to excess, threatens death, since the circulation of 'spiriti' in the body is essential to human life.

To summarize, Cavalcanti's depiction of the lyric subject's suffering in love through the metaphor of the automaton in sonnet VIII can be situated within a long-standing lyric ontology of love as death. The lethal depiction of love in Cavalcanti's oeuvre is rooted in the fact that love is understood as blocking the rational thought that defines human life as such. This inner death is also bound up with physical disruption, as evidenced through the motif of 'spiriti', mediators between body and soul according to medieval physiology and central agents of the 'death' of the lyric subject in Cavalcanti's oeuvre. Overall, the metaphor of the automaton in sonnet VIII is a particularly memorable instance of a more pervasive tendency throughout Cavalcanti's oeuvre and medieval lyric generally, to depict love as emptying the subject of inner forces, leaving him in an ontologically liminal state, on the brink of life and death.

Christ-as-Automaton: Religious Practice and Parody

The automaton metaphor is deployed by Cavalcanti in sonnet VIII precisely because of the uncanny mortal liminality that it conveys, offering a memorable image of the lyric subject's extreme suffering in love. The automaton, however, also functions as an inverse/perverse image of Christ, the liminally mortal figure par excellence of medieval Christian culture. Vecchi Galli (23) indicates that the automaton image of Cavalcanti's sonnet can simultaneously be read as a parody of Christ's Resurrection due to the detail of the wound in the heart ('porti ne lo core una ferita', *he bears in his heart a wound,* v. 13). The wound's location invites a Christological comparison since, from the twelfth century, Christ's side wound was understood specifically as a wound in his heart (Sauer 202–3). This reading of the automaton metaphor of sonnet VIII as a parody of Christ's Resurrection challenges the critical assumption that Cavalcanti's lyric oeuvre is paradigmatically 'secular', uninterested in religious themes. A contrast between Cavalcanti and Dante – between, as Stone puts it, 'a theological Dante versus a philosophical Cavalcanti' – is often introduced in support of this (*Guido* 1). While Stone's effort to counter the Dante-Cavalcanti dichotomy focuses exclusively on the presence of (often unorthodox) natural philosophical thought in Dante, this dichotomy might also be troubled by examining the converse possibility, unexamined by Stone, of religious imagery in Cavalcanti's oeuvre.

Ardizzone, in her book *The Young Dante,* like Stone, emphasizes the centrality of philosophy and natural science, including heterodox aspects, to Dante's works. However, the introduction to her book on Cavalcanti invokes the typical opposition between Dante-as-theological and

Cavalcanti-as-philosophical (*Guido* 6). This framework is instrumental in allowing Ardizzone, throughout the rest of her book, to elucidate the religious heterodoxy of Cavalcanti's thought, particularly given the contemporary controversy of Averroism. However, it somewhat obscures the fact that Cavalcanti's apparent heterodoxy does not necessarily exclude the presence of religious imagery, at least from an ironic or parodic perspective, in his oeuvre. Indeed, a minority of Cavalcanti scholars – most notably Paola Nasti and Ronald Martinez – have raised this suggestion. These scholars particularly focus on Cavalcanti's use of biblical citation, with Nasti examining the poet's engagement with the biblical Song of Songs and Martinez considering his deployment of citations from the book of Lamentations ('Man'). Elsewhere, Martinez has considered Cavalcanti's use of the Virgin Mary as a figure of his lady, with idolatrous implications ('Specter'). Nasti suggests that Cavalcanti's parodic intent in his deployment of such religious imagery and rhetoric is visible particularly in his stance on suffering: while, in medieval Christianity, pain was perceived as potentially productive and redemptive, enabling the suffering subject to purify themselves and imitate Christ, for Cavalcanti pain is ultimately nonproductive (Nasti 87). Cavalcanti's imagery of the automaton in sonnet VIII exemplifies precisely this stance. In contrast to Christ's Resurrection, for Cavalcanti, death triumphs: although the subject's death is repeated throughout his lyrics, true resurrection is impossible, as apparent revivals serve only as preludes for further deaths. As Battesti comments: 'this results in the reduction of narrative possibilities to a single choice: the death of the lover is no longer optional' (119).[14]

Martinez's analysis of Cavalcanti's usage, throughout his lyrics, of a particular verse from Lamentations (1.12) suggests a parallel for the Cavalcantian lyric subject in contemporary iconographical traditions.[15] As Martinez notes, this Biblical verse was linked to a particular image type: that of the 'Man of Sorrows', a representation of Christ as simultaneously dead and alive, encapsulating the Passion and Resurrection into a single frame ('Man' 197–9). Although Martinez mentions this iconographical context, his focus is on Cavalcanti's engagement with the biblical intertext rather than the visual tradition to which this biblical verse was connected. The neglect of the 'Man of Sorrows' as a visual intertext for Cavalcanti may be due to debates over the date of the arrival of the image type into Italy (Ridderbos 145–9). However, at least one example, an Umbrian Man of Sorrows, is datable to the mid-thirteenth century (1255–60), so such image types were present in Cavalcanti's Italy (Cannon). As such, the visual tradition of the 'Man of Sorrows' may have been an explicit referent for Cavalcanti, and, in any case, there are clear parallels with his lyric subject. Like the automaton, the 'Man of Sorrows' offers an image of mortal undecidability, figuring Christ as alive and dead at once. While, in the case of the automaton, this may provoke a sense of eeriness or horror, the intended affective responses to liminal images of Christ are awe and wonder, conveying to viewers the capacity of the divine to transgress seemingly natural ontological boundaries between life and death.

It may seem far-fetched to compare the life/death liminality embodied by Christ to that of the automaton, but besides Cavalcanti's sonnet, there is evidence that medieval Christians perceived this parallel. As already mentioned, the robotic animals built around 1300 at Hesdin are usually named as the first automata constructed in the Latin West. However, Christopher Swift proposes that the mid-thirteenth-century Spanish 'Virgen de los Reyes', a pair of articulating wooden statues of the Virgin and Child with internal mechanisms enabling autonomous movement of their heads, is the first such example. These statues, paraded into the defeated Islamic city of Seville by Fernando III of Castile in 1248 and subsequently owned by Alfonso X, are extant today in Seville Cathedral ('Technology'; 'Robot Saints'). Though constructed in Christian Spain, the 'Virgen de los Reyes' is unlikely to be cited as the first Western automaton due to Spain's complex geopolitical position between Christian and Muslim worlds in this period.

However, similar devotional objects such as wooden statues of Christ with articulating limbs were constructed and utilized in (para-)liturgical celebrations elsewhere in Europe from the thirteenth century onwards, particularly in Umbria and Tuscany (Kopania; Bernardi and Bini). Swift notes that such objects function differently to the 'Virgen de los Reyes' since they do not move autonomously but require puppeteers ('Robot Saints' 55–6). Yet these articulating figures are, like automata, craft objects made of wood or other inanimate materials that are intended to realistically mimic a human body through movement. Additionally, self-moving automata like the 'Virgen de los Reyes' were not necessarily clearly conceptually distinguished even from apparently immobile statues and paintings, and, as Swift writes, 'a medieval metaphysics of matter allows for inert statues and automata to belong to the same intuitive domain' ('Technology' 34). Occasionally, statues and paintings of holy figures were thought to come alive and perform miracles, as was the case with the 'Orsanmichele Madonna' in Florence, an image of the Virgin Mary that, during the 1290s, became known for its miracle-working abilities, as the chronicler Giovanni Villani reports:

> Great and public miracles began to be demonstrated in the city of Florence thanks to a painted figure of the Virgin Mary on a column in the Orto Sanmichele *loggia* where the grain market takes place, curing the sick, restoring movement to the paralysed, and visibly freeing a great number of people possessed by demons.[16]
>
> (628)

One of Cavalcanti's sonnets (XLVIII), addressed to fellow poet Guido Orlandi, refers explicitly to this image, reporting in similar terms to Villani the various miracles performed by it.[17] Caroline Walker Bynum situates such miracle-working images within a complex, contradictory ontology of matter endemic to late medieval Christianity (105–121). Matter was, by definition, living, animate 'stuff', subject to change and decay: 'the basic way of describing matter [...] was to see it as organic, fertile, and in some sense alive' (Bynum 30). The potential 'liveliness' of art objects thus did not depend on their naturalism or realism but was inherent in their objectness (Bynum 125). However, the dynamic mode of life attributed to matter coexisted, for medieval Christians, with a different, and superior, definition of 'life'. True life was located not in matter but in God's eternity, characterized by a permanence and unchangeability that, from the perspective of organic life, is not very lively at all. Bynum calls this the 'essential paradox of eternal life': 'life, which can be understood only by analogy to the organic and thus changeable must, if eternal, never undergo the generation and corruption that, however, defines it' (57).

The automaton embodies these contradictions inherent to medieval Christian ontologies of matter. The humanoid automaton offers a representation of the human body if it were reduced to mere matter, deprived of the rational soul that renders it potentially eternal. Yet the durability of the materials of which automata are made also seems to prefigure the eternal post-Resurrection body: 'robots are gestures towards immortality, created in the face of the undeniable fact and experience of the ongoing decay of our fleshly bodies' (Swift, 'Robot Saints' 52). If the matter of which the automaton is constructed is not eternal, it nonetheless gestures in this direction by being more durable than the organic matter of the corpse. In the religious context, it is this interpretation of the automaton, its capacity to point towards an eternal mode of embodiment, that renders it devotionally effective and Christ-like. As Swift suggests, the 'domain-specificity' of religion 'could shield the spectator of an enigmatic devotional android from the sense of the uncanny' ('Robot Saints' 59). That is, the devotional automaton (or articulating statue or miraculous image) does not evoke eeriness or horror, since Christianity's promise of eternal life makes a claim that already breaches the ontological boundaries of life and death, but in a way that does not threaten human life but crystallizes it in eternity. However, Cavalcanti's imagery, while alluding to the Christ-like automaton, does not 'shield' the spectator in this way but parodically reverses the directionality of such Christian imagery to point not towards eternal life but eternal death, thus reintroducing the sense of the 'uncanny'.

In their capacity to foreshadow eternal, heavenly modes of embodiment, the conceptual function of automata in religious culture may be perceived as parallel to saintly relics. Indeed, Truitt gestures towards the 'conceptual affinity' between automata and relics, since the latter, resistant to the decay that usually afflicts corpses, exist in an 'in-between state' that 'articulates the boundary work that medieval automata [also] undertake' (*Medieval* 97). The phenomenon of incorrupt saintly corpses or body parts, many of which also demonstrated their 'liveliness' by performing healing miracles or exuding pleasant odours, was a common late medieval miracle, asserting the saint's triumph over death in eternal life (Vauchez 427–43; Bynum 177–92). The incorruptibility of saintly relics was enhanced by their placement in reliquaries in the shape of body parts but made of durable materials like metals and gems, gesturing towards the impermeability of the glorified post-Resurrection body (Bynum 70–1, 132, 184–5). The resistance of relics to bodily change and decay, enhanced by the durable materiality of reliquaries, endows saintly bodies with a mode of 'liveness' beyond the organic, more closely resembling the apparent permanence of the corporeal life of the automaton.

By comparing his lyric subject simultaneously to an automaton and to Christ, Cavalcanti gestures towards the complexities and ambivalences characterizing medieval Christian ontologies of matter and embodiment. Cavalcanti's metaphor is an uncanny, horrifying depiction of a human body reduced to matter alone, deprived of the rational soul and inner life that, for medieval philosophers, constituted the nature of human existence. Yet the parallel implicitly evoked by the sonnet between Christ's mortal liminality and that of the automaton seems to gesture – as do the articulating Christ statues, miracle-working images, and reliquaries of medieval devotional culture – towards an eternal mode of life, removed from the transience and decay that characterizes organic existence. The temporal existence of Cavalcanti's lyric subject could indeed be perceived as approximating eternity, insofar as lyric is – as Culler emphasizes – characterized by 'iterability', as 'a moment of time that is repeated every time the poem is read' (295). However, if the Cavalcantian lyric subject exists in a kind of 'eternal present' in the space of the lyric, this is not the eternal life promised by Christ's Resurrection, but an eternal death. By enacting ad infinitum the flight of the lyric subject's 'anima' or 'spiriti' and the consequent death of his inner self every time the lyric is read, Cavalcanti parodies and demystifies the use of the automaton as a conceptual analogy for Christ's Resurrection, bringing back the sense of the uncanny that is defused in such religious usage of automata.

Conclusion

The automaton, in medieval European literature, functioned as a powerful conceptual tool, even if actual physical automata remained rare and mysterious, at the margins of the Western world. The literary automaton functioned to delineate and trouble ontological binaries: between natural and artificial, licit and illicit knowledge, life and death. It is specifically the automaton's transgression of categories of alive/dead on which Cavalcanti draws when he evokes this figure in sonnet VIII. Cavalcanti's automaton metaphor may be situated in the context of a long lyric tradition of representing love as living death, perpetuated by the troubadours and inherited by the early Italian lyric tradition. Cavalcanti's lyric oeuvre particularly accentuates this motif, consistently representing the loving subject as undergoing a death of the inner rational activity that defines the human.

However, the Cavalcantian automaton also functions as a parody of the Resurrection, as an inversion or distortion of the mortal liminality represented by Christ. Various devotional artefacts suggest that medieval Christians perceived the parallel between the automaton and Christ. Yet, in the religious context, the automaton does not represent an uncanny human body deprived of inner life so much as it represents a more durable form of embodiment, foreshadowing the post-Resurrection body. Despite alluding to a Christological parallel alongside his imagery of the automaton, Cavalcanti discards the positive, hopeful implications of the devotional function of the automaton as a symbol of eternal life. The Cavalcantian automaton inverts the eschatological significance of the automaton in religious contexts by embodying not eternal life but eternal death, as the flight of his 'spiriti' or 'anima' is performed repeatedly in the circularity of the lyric. We are faced with a horrifying spectacle of a human body reduced for eternity to a hollow shell, deprived of the agency, sentience and intellectual activity that, for thinkers across history, has seemed crucial to distinguishing human from machine.

Notes

1 'È con lui che inizia la lirica moderna in senso stretto, in quanto indagine poetica, autocosciente dell'*homo interior*'. All English translations given throughout this chapter are my own.
2 All citations from Guido Cavalcanti's poems are from Roberto Rea and Giorgio Inglese's edition.
3 'Artificio (del maestro che l'ha fabbricato), congegno, meccanismo'.
4 For instance: 'La condizione descritta dal paragone con l'automa semovente [...] arriva a postulare uno stato di cronica alienazione da sé' (*The condition described by the comparison with the self-moving automaton [...] functions to postulate a state of chronic self-alienation.* Rea in Cavalcanti, *Rime* 72); 'La condizione dell'automa cancella cioè ogni traccia di vita effettiva' (*The condition of the automaton erases, therefore, any real trace of life.* De Robertis in Cavalcanti, *Rime* 29).
5 Quotation from the anthology of troubadour poetry edited by Carl Appel.
6 Quotation from Ernest F. Langley's edition of Giacomo da Lentini's poetry.
7 Quotation from Luciano Rossi's edition of Guido Guinizzelli's *Rime*.
8 'Un vero *hapax* nella poesia del tempo di Dante'.

9 Quotation from Francesco Egidi's edition of Guittone d'Arezzo's *Rime*. Guittone's sonnet is identified as a possible source for Cavalcanti by Rea (in Cavalcanti, *Rime* 74) and Rigo (124–5).
10 Citations from Dante Alighieri's *Convivio* from Franca Brambilla Ageno's edition.
11 '"Morte spirituale", cioè interruzione di quella attività intellettuale che è sola vera *vita* dell'uomo'.
12 'Dans le corpus lyrique de Guido, cette non-vie ontologique ne se distingue pas toujours nettement de la mort physique'.
13 See, for instance: XIII. 1–6, IX. 9–14 (pp. 97–8 and 78 in Rea and Inglese's edition).
14 'Il en résulte [...] la réduction des possibles romanesques à un seul: la mort de l'amant n'est plus optionnelle'.
15 The most obvious citation of Lamentations 1.12 in Guido's oeuvre is XIX.1–3 (pp. 117–18 in Rea and Inglese's edition).
16 'Si cominciarono a mostrare grandi e aperti miracoli nella città di Firenze per una figura dipinta di santa Maria in uno pilastro della loggia d'Orto Sammichele, ove si vende il grano, sanando infermi, e rizzando attratti, e isgombrare impersati visibilemente in grande quantità'.
17 See pp. 251–3 of Rea and Inglese's edition for the full text of this sonnet.

References

Abulafia, David. *Fredrick II: A Medieval Emperor*. Pimlico, 1992.

Alighieri, Dante. *Convivio,* I. Introduzione (Tomi I e II), II. Testo (Tomo III), a cura di Franca Brambilla Ageno. Le Lettere, 1995 (Le Opere di Dante Alighieri. Edizione Nazionale a cura della Società Dantesca Italiana, 3) [Edizione riveduta della Società Dantesca italiana].

Alighieri, Dante. *Inferno*. Edited by Anna Maria Chiavacci Leonardi, Oscar Mondadori, 2005.

Antonelli, Roberto. "Cavalcanti o dell'interiorità." *Alle origini dell'io Lirico*, edited by Roberto Antonelli, Viella, 2001, pp. 1–22.

Appel, Carl. *Provenzalische inedita aus Pariser Handschriften*. G. R. Reisland, 1892.

Ardizzone, Maria Luisa. *Guido Cavalcanti: The Other Middle Ages*. U Toronto P, 2002.

Ardizzone, Maria Luisa. *The Young Dante: Archetypes of His Early Intellectual Biography*. Cambridge Scholars Publishing, 2022.

Aristotle, *De motu animalium*. Translated by Martha Nussbaum, Princeton UP, 1978.

Barolini, Teodolinda. *Dante and the Origins of Italian Literary Culture*. Fordham UP, 2006.

Battesti, Isabelle. "Temporalités de la mort amoureuse dans la poésie de Guido Guinizzelli et Guido Cavalcanti." *Les deux Guidi, Guinizzelli et Cavalcanti*, edited by Marina Gagliano, Philippe Guérin, and Raffaella Zanni, Presses Sorbonne Nouvelle, 2016, pp. 107–22.

Bernardi, Claudio, and Carla Bini. "Interactive Images: Depositions of Christ in Italy 1200-1700." *Good Friday Ceremonies with Articulated Figures in Medieval and Early Modern Europe*, edited by Christophe Chauguinian, Classiques Garnier, 2023, pp. 233–309.

Brumbaugh, R. S. *Ancient Greek Gadgets and Machines*. Cromwell, 1966.

Bynum, Caroline Walker. *Christian Materiality: An Essay on Religion in Late Medieval Europe*. Zone Books, 2011.

Caciola, Nancy Mandeville. *Afterlives: The Return of the Dead in the Middle Ages*. Cornell UP, 2016.

Cannon, Joanna. "The Stoclet 'Man of Sorrows': A Thirteenth-Century Italian Diptych Reunited." *The Burlington Magazine*, vol. 141, no. 1151, 1999, pp. 107–12.

Cavalcanti, Guido. *Rime, con le Rime di Iacopo Cavalcanti*. Edited by Domenico De Robertis, Einaudi, 1986.

Cavalcanti, Guido. *Rime*. Edited by Roberto Rea and Giorgio Inglese, Carocci Editore, 2011.

Ciavolella, Massimo. "L'amore e la medicina medievale." *Guido Cavalcanti tra i suoi lettori*, edited by Maria Luisa Ardizzone, Cadmo, 2003, pp. 93–102.

Corti, Maria. *La felicità mentale: nuove prospettive per Cavalcanti e Dante*. Einaudi, 1983.

Culler, Jonathan. *Theory of Lyric*. Harvard UP, 2015.

Da Lentini, Giacomo. *The Poetry of Giacomo da Lentini: Sicilian Poet of the Thirteenth Century*. Edited by Ernest F. Langley, Harvard UP, 1915.

D'Arezzo, Guittone. *Le Rime*. Edited by Francesco Egidi, Laterza, 1940.

Daston, Lorraine, and Katharine Park. *Wonders and the Order of Nature, 1150–1750*. Zone Books, 1998.

Descartes, René. "Treatise on Man." *The Philosophical Writings of Descartes*, edited and translated by John Cottingham, Robert Stoothoff, and Dugald Murdoch, vol. 1, Cambridge UP, 1984, pp. 99–108.

Farmer, Sharon. "La Zisa/Gloriette: Cultural Interaction and the Architecture of Repose in Medieval Sicily, France and Britain." *Journal of the British Archaeological Association*, vol. 166, no.1, 2013, pp. 99–123.

Guinizzelli, Guido. *Rime*. Edited by Luciano Rossi, Einaudi, 2002.

Kang, Minsoo. *Sublime Dreams of Living Machines: The Automaton in the European Imagination*. Harvard UP, 2011.

Kopania, Kamil. "Animated Sculptures of the Crucified Christ: Origins, Development, and Impact." *Material Religion*, vol. 14, no. 4, 2018, pp. 545–58.

MacDorman, Karl F., and Hiroshi Ishiguro. "The Uncanny Advantage of Using Androids in Cognitive and Social Science Research." *Interaction Studies*, vol. 7, no. 3, 2006, pp. 297–337.

Martinez, Ronald L. "Cavalcanti 'Man of Sorrows' and Dante." *Guido Cavalcanti tra i suoi lettori*, edited by Maria Luisa Ardizzone, Cadmo, 2003. 187–212.

Martinez, Ronald L. "Guido Cavalcanti's 'Una figura della donna mia' and the Specter of Idolatry Haunting the Stilnovo." *Exemplaria*, vol. 15, no. 2, 2003, pp. 297–324.

Mayor, Adrienne. *Gods and Robots: Myths, Machines, and Ancient Dreams of Technology*. Princeton UP, 2018.

Mori, Masahiro, "The Uncanny Valley." Translated by Karl F. MacDorman and Norri Kageki, *IEEE Robotics and Automation Magazine*, vol. 19, no. 2, 2012, pp. 98–100.

Nardi, Bruno. "L'averrismo del 'primo amico' di Dante." *Studi Danteschi*, vol. 25, 1940, pp. 43–79.

Nasti, Paola. "Nozze e vedovanza: dinamiche dell'appropriazione biblica in Dante e Cavalcanti." *Tenzone: Revista de la Asociación Complutense de Dantología*, vol. 7, no. 1, 2007, pp. 71–110.

Pattison, Joel. "A Golden Tree in the 'Garden of Pages': The Genoese Embassy to Morocco of 1292." *Journal of Medieval Worlds*, vol. 1, no. 4, 2019, pp. 1–9.

Ridderbos, Bernard. "The Man of Sorrows: Pictorial Images and Metaphorical Statements." *The Broken Body: Passion Devotion in Late Medieval Culture*, edited by A. A. MacDonald, H. N. B. Ridderbos, and R. M Schlusemann, Egbert Forsten, 1998, pp. 145–81.
Rigo, Paolo. "'Io vo come colui ch'è fuor di vita': un topos letterario del Duecento." *La poesia in Italia prima di Dante*, edited by Franco Suitner, Longo Editore, 2017, pp. 115–29.
Sauer, Michelle M. "Queer Time and Lesbian Temporality in Medieval Women's Encounters with the Side Wound." *Medieval Futurity: Essays for the Future of a Queer Medieval Studies*, edited by Will Rodgers and Christopher Michael Roman, De Gruyter, 2021, pp. 199–219.
Stone, Gregory. "Animals are from Venus, Human Beings from Mars: Averroës' Aristotle and the Rationality of Emotion in Guido Cavalcanti's 'Donna me prega'." *PMLA*, vol. 130, no. 5, 2015, pp. 1269–84.
Stone, Gregory. *Guido Cavalcanti: Poet of the Rational Animal.* Routledge, 2020.
Swift, Christopher. "Robot Saints." *Preternature*, vol. 4, no.1, 2015, pp. 52–77.
Swift, Christopher. "Technology and Wonder in Thirteenth Century Iberia." *Performing Objects and Theatrical Things*, edited by Marlis Schweitzer and Joanna Zerdy, Palgrave Macmillan, 2014. 21–35.
Truitt, Elly R. "Fictions of Life and Death: Tomb Automata in Medieval Romance." *Postmedieval*, vol. 1, 2010, pp. 194–8.
Truitt, Elly R. *Medieval Robots: Mechanisms, Art, Magic.* U Pennsylvania P, 2015.
Truitt, Elly R. "'Trei poëte, sages doctors, qui mout sorent di nigromance': Knowledge and Automata in Twelfth-Century French Literature." *Configurations*, vol. 12, 2005, pp. 167–93.
Vauchez, André. *Sainthood in the Middle Ages.* Translated by Jean Birrell, Cambridge UP, 2005.
Vecchi Galli, Paola. "L'automa da Cavalcanti a Dante." *Da Dante al Novecento: in onore di Alfredo Cottignoli*, edited by Sebastiana Nobili et al., Pàtron Editore, 2014, pp. 19–28.
Villani, Giovanni. *Nuova Cronica.* Edited by Giuseppe Porta, vol. 1. Guanda, 1990.
Webb, Heather. *The Medieval Heart.* Yale UP, 2010.

4 Monsters, Mechanics and Automatic Writing in E.T.A. Hoffman's 'The Sandman' and Gérard de Nerval's 'Aurélia'

Vanessa C. Weller

Introduction

Autonomous movement of inanimate objects, for centuries deemed the work of ghosts and demons, has, since the Industrial Revolution, become the provenance of mechanical machinery and a facet of everyday life. With the advent of Artificial Intelligence and supercomputers in the late twentieth and early twenty-first centuries, even cerebral tasks that were once considered solely the domain of human beings have been exported to machines and been automated. This is both an advantageous convenience and a frightening, brave new world. Yet humanity's ambivalent relationship to technology is not new. In the late eighteenth and early nineteenth centuries, amid vast scientific advances and untold potential, great instruments of progress and instruments of devastation became reality. At the precipice of modernity, when industrialization, capitalism and urbanization overtook rural life for many in Europe, the uneasiness of accepting this new mode of existence which separated humanity from the natural world more brutally than ever before spurred philosophical treatises and literary movements into existence.

Literary hypotheses of the science fiction and fantasy genre allowed a comfortable distance from otherwise psychologically overwhelming concepts. The discussion of automatons and automation particularly invited the consideration of what folklore had previously deemed the work of otherworldly – and often malevolent – beings, and what Freud would later term the Uncanny. Gothic themes, embraced at this time in France and Germany, bring to light nascent fears rooted in folklore and propelled by the unknowns of technological advancement. The Ur-robotics of E.T.A. Hoffmann's *The Sandman* (1817) explores both the legend of the doppelgänger and the psychical fallout of developing an uncanny human-like presence that is, despite appearances, uncompromisingly mechanical. Redefined by Hoffmann's French inheritor, Gérard de Nerval, in his autobiographical novella *Aurélia* (1855), automation takes the form of automatic writing, where the author's very real psychotic breaks of 1841 and

DOI: 10.4324/9781003545910-5

1852 are mirrored in fiction, positing the antagonist as the narrator's own subconscious, and blurring the lines between the protagonist's imaginary and the workings of the universe.

This chapter analyses how automation, medicine and consciousness interact in these literary works that straddle the divide between folklore and the science fiction and fantasy genres. The ultimate, godlike power of shaping another in one's own image is at the crux of these works. Each protagonist falls in love with an unattainable female presence which leads to his psychosis. Through communion with both the technological and the otherworldly, fantasies of creating or becoming the perfect romantic partner, ordinarily out of reach to mere mortals, are made possible for Hoffmann's Nathanael and the unnamed narrator of *Aurélia*. The fractious nature of changing oneself, or forcing change upon another, dehumanizes the protagonist of each novella, as he, in turn, strips the humanity from his object of affection. What follows is an investigation of humanity's relationship to nature via technology, the dark side of advancement which necessarily includes the burden of gender roles within this relationship, and an interrogation of how far our society, still burdened by existential nightmares, has truly come.

The Pygmalion Precedent

The metaphor of humanity's dual nature, rooted in antiquity, is ubiquitous in the works of the Romantics. Duality here is ultimately a stand-in for moral decision-making when one is at a crossroads. Goodness and malevolence, or morality and immorality, find symbolic resonance in technological innovation. The inventor oftentimes chooses to create in order to improve humanity's lot, and the unintended consequences of bringing the invention into the world – a storyline so prevalent in tales of science fiction – are just as often framed as hubris on the part of the inventor. Modernity creates duality in the sense that contemplation of the past – folk wisdom and the pastoral – is a natural reaction to the rush of technological innovation which, because of the rapidity of change, creates feelings of alienation in those who are most vulnerable to it. Seemingly unstoppable movement towards progress, combined with analytical ways of viewing the physical world – such as mechanized labour and Freudian psychoanalysis – engender a fear of humanity being replaced by machines, precisely because the analytic (mechanical) subsumes the emotional (human). Such dualities are symptomatic of civilization's innate need for moral structure, itself mirrored in both the theological and the poetical. From the dichotomous terms of 'good' and 'evil' to the more complex Freudian *ego*/*id* and Jungian 'self'/'shadow', the ancient struggle for balance presents in both self-conception and interpersonal relationships.[1]

Plato's *Symposium* is likely the most famous source of the myth of the bifurcation of the 'soul mate'. In *The Symposium*, the playwright Aristophanes claims that this primordial separation, that is, the creation of male and female from formerly hybrid beings, was punishment for humanity achieving a sort of perfection to rival the gods. Zeus, fearing that these perfect humans might revolt, split them in two, and from this comes this existence of soulmates. To be truly loved, one must find one's 'second half'. The empiricism of the Enlightenment pushed back on this idealized notion of love, '[restoring] us to our original state'.[2] Platonism's relevance to our contemporary world, then, is due in large part to Romanticism (Hedley 270). The embrace of creative and artistic depictions of nature as expression rather than mimesis, the interpretation of both the natural and supernatural worlds, in a sense returns the ideals of the nineteenth century to Platonism. This can be seen in further examples of idealized (yet taboo) love originating in antiquity.

Ovid's *Metamorphoses* introduces Pygmalion, a sculptor who creates an ivory statue of the perfect woman, called Galatea in some versions of the myth. Venus, the goddess of love, brings Galatea to life in answer to Pygmalion's prayers. In Apuleius's *The Golden Ass*, Cupid, in love with the beautiful yet mortal Psyche, disguises himself to seduce her. As Psyche lives in darkness, unaware of the identity of her lover, her curiosity gets the better of her, and she surreptitiously discovers his identity. As punishment, Psyche is forced to walk the earth looking for Cupid, with a trip to the underworld the culmination of her trials. These antique precedents for the sexualized manipulation of inanimate objects or lesser-beings-as-love-interests are echoed in Hoffmann and Nerval and appear in myriad guises throughout the Romantic movement. The soul, the mind and the concept of free will were central to the thesis of Romanticism. Thus, a preoccupation with such themes, along with the Gothic, an exploration of the dark side of the natural order – the daemonic as opposed to the sublime – taps into the philosophical framework of moral decision-making, reevaluating its motives in opposition to Enlightenment ideals. In the early twentieth century, psychoanalysis would famously take up the mantle of inward and outward explanations of the human psyche to explain what to many had previously appeared inexplicable. The alienation caused by societal progress is explained by the psychoanalytic theories of Freud and Jung. Freud's preoccupation with the interior (how individuals confront their reality) and Jung's emphasis on the exterior (how society at large deals with such decisions) effectively attempt to explain the mind's process of rationalization. These theories counter and reflect upon the overwhelm engendered through the innovations of modernity. But before psychoanalysis, literature took up the mantle of piecing together human nature through disparate realities.

Monsters and Mechanical Women

In the German literary tradition, the metaphor of the double, or *der Doppelgänger,* situates the duality of human nature in a malevolent force typically taking the identical form of a living being and portending the original's death.[3] Clearly defined by E.T.A. Hoffmann in *Die Elixire des Teufels* (*The Devil's Elixirs*, 1815) and further explored in *Der Sandmann* (*The Sandman*, 1817), the double here is an undoubtedly evil presence. In the form of antagonist Coppelius/Coppola, Hoffmann takes the Sandman, typically a benign character in German folklore associated with children's bedtime and the rheum or 'sand' that accumulates on the eyes during sleep, and turns him into a monster. Friedrich Gottlieb Klopstock (1724–1803) had dealt with similar esoteric and mythical themes in poetry a generation prior, and Goethe (1742–1832) and Schiller (1759–1805) revisited these themes in poetry and theatre.[4] Thus, *The Sandman's* innovation was not to invoke the phantasmagorical in writing, but the capacity to push the perception of the real into the unreal, to reconstitute the (super)natural world as a monstrous funhouse mirror:

> Hoffmann was the great innovator of a genre that had not varied its formula since the Middle Ages, even in *Faust*. He refined the fantastic by separating it from the marvellous. After the lovely expression of Barbey d'Aurevilly, he obeyed 'a voice who called from the beyond of being,' but he took pains, at least on his good days, to stop just before he got to the supernatural, at the very limits of the possible, if not of the real.[5]
> (Barine 3–4)

The malevolent presence, known to audiences as a stock character of medieval morality plays, took on a further incarnation in the character of Mephistopheles in Goethe's *Faust*. Hoffmann strips him of his overt supernatural abilities, although a certain type of magic in the form of hypnosis is still present. The supernatural presence of the Mephistophelean devil becomes all the more sinister in *The Sandman* because it is banal and quotidian. Workaday solicitor Coppelius, business partner of Nathanael's father, is the everyday evil of avarice and disregard, as he opportunistically attempts to steal the father's share of the business profits. Furthermore, Coppelius either causes or does nothing to prevent the accident that kills Nathanael's father.

As a boy, Nathanael is encouraged by his nanny to believe in the supernatural, and he believes Coppelius to be the Sandman, whose distaste for naughty children descends into kidnap, torture and murder/cannibalism.

> He is a wicked man, who comes to children who refuse to go to bed, and throws a handful of sand into their eyes, so that they spring forth

> bleeding from their heads. He puts their eyes in a bag and carries them to the crescent moon to feed his own children, who nest up there. They have crooked beaks like owls so that they can peck out the eyes of naughty human children.[6]

The mystery surrounding the father's death initiates a threat to Nathanael's literal vision which slowly morphs into a metaphorical distortion. 'The Sandman brought me on the path of the marvellous, of adventure, that so easily nestles into a child's mind'.[7] His entire life becomes a nightmare, taking hold at the death of his father. When, during Nathanael's days as a university student, Coppelius returns as the optician Giuseppe Coppola, this childhood trauma resurfaces to impair Nathanael's adult life, manifesting in his relationship with women and woman-like facsimiles.

Hoffmann's reliance on the scientific is paramount to understanding how the unsettling nature of Coppelius/Coppola differs from prior incarnations of the malevolent force. Descartes' theory of the dualism of mind and body, that the mind is immaterial and ephemeral (existing beyond the physical world), whereas the body is both corporeal (grounded in the material world) and non-thinking, would have had equal sway over intellectuals at the beginning of the nineteenth century as the Neoplatonism of the Romantics. The eighteenth century was replete with new waves of sophisticated automatic technology. Pierre and Henri-Louis Jaquet-Droz, and Jacques de Vaucanson, among others, produced lifelike mechanical devices that imitated human movements and proclivities with the intent to decode what it meant to be human (Cheng 452). The humanoid automaton, although appearing human, does not have the range of emotions of a human being – or, arguably, any semblance of emotions – owing to the automaton's lack of what is generally known as a *soul. The Sandman* appears at the end of this period of technological innovation that, for the first time, allowed for a public conception of lifelike androids and their possible inner lives. Along with the burgeoning bourgeoisie, these fictionalized encounters predict a certain type of human reaction to the artificial intelligence currently debated in our time. Lin Cheng offers a theory on android emotions as being either simulated by the AI itself, as is the case in contemporary AI models such as ChatGPT, or as a projection of the emotions of the humans around the automatons, as is the case with Hoffmann's Olimpia (454).

Blindness to an objective reality obscures Nathanael's reasoning due to his childhood trauma. To direct his reality, which he feels is beyond his control, he falls in love with the humanoid automaton Olimpia, jilting his human fiancée, Clara, who, as her name suggests, symbolizes reason and clarity. She is the first to offer a logical answer to Nathanael's fears, telling him that the events he claims to have experienced are all in his head.

Nathanael is at first assuaged by Clara's words, but she, as the voice of reason in his life and a metonymical stand-in for an Enlightenment philosophy, tires of his mystical and metaphysical musings. He ultimately dismisses Clara's proto-psychoanalytical explanation to search for a spiritual explanation of his own.

Nathanael's rejection of Clara – and by extension her logic – feeds into the misogynist notion that women are irrational even when they demonstrate a rationality that, coming from a masculine voice, would be considered and evaluated on its merits. Furthermore, Clara exemplifies the overarching societal expectations of women in the late eighteenth and early nineteenth centuries, that they abstain from a public life and a life of the mind, and if they do not, they must be punished. Aurora B. Romero argues: 'Femininity is thus related to both a reduction of authenticity and a diminishment of knowledge that is directly associated with the appearance of women's bodies and their performance of gender' (66). Gender performance, such as that which is expected of Clara and of women of her social standing in this era, reduces both feminine authenticity and female bodily autonomy, ultimately denying a woman's intellectual potential in order to shoehorn her into the societally acceptable role of caretaker, depriving women of an aspect of their humanity. Olimpia, by contrast, is exhibited as a spectacle by her 'father' Professor Spalanzani, which is where Nathanael encounters her and falls in love. Butlerian gender-as-performance here morphs into gender-as-technology. Just as in Beth Coleman's 2009 argument in 'Race as Technology' that race must be distanced from the biological and thought of in terms of its usefulness, gender could also be used as a tool, external to identity. By separating race – and, I argue, gender – from the genetic and thus predetermined circumstances of human existence, agency becomes possible.[8] The preset, prescriptive concept of 'female' or indeed 'womanhood' is weaponized: in Olimpia's case, to cause chaos, and in Clara's case, as a self-defence mechanism. By going against Nathanael and enacting her own agency, Clara refuses his version of reality – and of patriarchy – and thus renounces a happy ending with him.

In this light, the appeal of Olimpia to Nathanael appears more understandable. Unlike Clara, Olimpia will not oppose him, find him boring or even have ideas of her own. She acts not only as an idealized woman but also as a force majeure in Nathanael's life, further separating him from an objective reality. Clara destroys his self-esteem and the vision he maintains of his Self, due to her own agency and free will. Even if Clara's goal is to help Nathanael and to love him, she is not ready to deny reality or her perception of facts for his well-being, which he demands (by negation) when he lobs the ironic insult accusing Clara of being a 'lifeless automaton'.[9] Clara cannot give Nathanael the reassurance that his interior Self requires. He produces a subjective reflection of the world that is ultimately unreal.

He is thus narcissistically attracted by the image of Olimpia, an automaton which becomes not only complicit in his warped worldview, unable to offer her own alternative, but also the literal *object* of Nathanael's affection. The conviction of his own truth and perspective keeps Nathanael in a precarious state of health. Clara's rejection of Nathanael's truth is worse than any possible romantic rejection. He prefers to invent a woman created by his own perceptions, just as he creates an enemy in Coppelius/Coppola as the Sandman. The discovery of Olimpia's non-existence – that she is an automaton – is the event that finally destroys him. Clara's escape from Nathanael and her ultimate happy ending – married to someone else – arguably points to a proto-feminist message from Hoffmann.

Metaphysics and Psychoanalysis

It is in this context of fluidity between the real and the unreal which so disturbs Nathanael that the narrator of *Aurélia* also finds himself. The narrator's disturbances, described in a fluid, poetic voice in *Aurélia*, reflect Nerval's real-life crises in 1841 and 1852. Part hallucination, part dream, the narrator shares his experiences to soothe his anxiety, just as Nathanael's letters to Lothar and Clara are his attempt to comprehend the events around the death of his father. However, Nerval's narrator finds the guilty party in his own subconscious – rather than in an external evil such as Coppelius – and it is in the dream world where he encounters his double.

The notable Germanic influence of the fantastic on Nerval's own writing is not unexpected. Nerval, accomplished as the French translator of Goethe, Heine and Klopstock, embraced the work of the German Romantics and allowed them to influence his own writing. Although Nerval likely needed no translation, several French versions of Hoffmann's work existed when *Aurélia* was published in 1855.[10] The prevalence of imported fantastical stories of both European and American provenance – such as the work of Edgar Allan Poe – points to the acceptance by the French public of the Gothic. Through the advocacy of Théophile Gautier, Charles Baudelaire and later Stéphane Mallarmé, French authors would adapt Gothic sensibilities so that, by the end of the century, fantasy fiction had become a thoroughly Gallic cultural product.

A distinct pattern emerges in the grotesque themes that originated in Germany when considering the shift from Goethe to Hoffmann to Poe to Nerval. Craig E. Stephenson, in his introduction to Jung's 1942 and 1945 lectures on *Aurélia*, describes the literary precedents of a descent into the underworld that so fascinated Nerval – Apuleius, Dante, William Blake – as part of a Christian cosmology. Goethe's 'cosmology of a nonbeliever' (Stephenson 14) arguably exerts the greatest influence on Nerval's work and the rejection of a Christian god and thus of Christian demonology. Nerval's descents into madness 'feel individual' (14) precisely because he is chronicling his own dissociation from reality through the conceit of narrative – his personal hell. Nerval's narrator claims a sort of paganist set of superstitions that seems at once atheist and pantheist, nevertheless preventing him from reaching heaven in the afterlife (Nerval 322). Seeking a spiritual explanation for his irrational ideas and disturbing visions, Nerval's unnamed narrator fights the malevolent forces that live inside him. The doppelgänger in *Aurélia* is at once the shadow of the narrator and an otherworldly power.

If the malevolent force terrorizing the narrator of *Aurélia* is part of the Self, it is not only impossible to differentiate between the narrator and his double, but, equally important, to separate them. Thus, all supernatural abilities granted to the doppelgänger also exist in the narrator, and the otherworldly visions that so disturb him also grant him a certain strength. Nerval commingles the limits of reality and fantasy in such an effective way that he has been accused of pastiche, borrowing the themes of *Aurélia* – the double and the eternal feminine – from Hoffmann (Warren 134). This false reading of Nerval is demonstrable when considering both Nerval's writing style and the autobiographical elements of his works which reveal a sympathy that is ultimately impossible in *The Sandman*. Hoffmann deals with Nathanael's traumas in a cold, analytic manner, in contrast to those of the narrator of *Aurélia*, which are lively, visceral and terrifying due to their proximity to the real. The Uncanny remains squarely in the realm of fiction with Hoffmann. Nathanael's hallucinations and breaks with reality are continually rationalized – both by Clara and Nathanael, and even as the scientific progress of Professor Spalanzani – up until Nathanael's suicide. Olimpia, like Coppelius/Coppola, is the projection of an unstable mind and a symptom of mental illness. Nerval, on the contrary, insists that his narrator's dreams and hallucinations are signs and synchronicities, proof of an alternate reality, albeit one that only he can see. The narrator's double is a true menace, terrorizing him from and within another realm. Additionally, unlike Nathanael, Nerval's narrator refuses to accept his psychosis.

Nerval's choice to deal with personal, biographical details in the manner of what would become known as autofiction in the twentieth century complicates authorial voice and reader perception. The narratological 'autobiographical pact' of Lejeune presumes that the identity of the narrator is equal to that of the author and that the author's 'honourable intention' is to share a demonstrably truthful and biographical narrative (El Bahi 306–7). Although Hoffmann positions the events of *The Sandman* as possible, if not plausible, in the opening epistle of the novella, he by no means posits the events as 'real' in the way that Nerval does through the use of elements of what would later be termed the genre of autobiography, or, in some instances, autofiction. This genre has proven to resonate with a French reading public, who arguably encountered it as early as in Rousseau's *Confessions* (1782). Nerval's poetic licence and his personal correspondence linking the events of *Aurélia* to his own life events and in particular his mental crisis and 1852 hospitalization (Mizuno 334) allow for a fluidity of interpretation.

Nerval's narrator suffers from a romantic rejection exacerbated by the death of Aurélia, a woman he has loved from afar for many years. Just as the discovery of Olimpia's true nature destroys Nathanael, the trauma of an amorous rejection is the inciting incident of the mental and rational decline of Nerval's narrator. Seeing her at a gathering and feeling compelled to speak to her, he apologizes for his past behaviour and receives Aurélia's forgiveness (Nerval 294). Then, a few days later, the narrator experiences his first crisis when he sees Aurélia's doppelgänger:

> One evening, about midnight, I was walking up an isolated street toward my lodgings when, glancing up by chance, I noticed the house number illuminated by a streetlamp. The number was that of my age. Immediately, lowering my eyes, I saw before me a pale, hollow-eyed woman who appeared to resemble Aurélia. I said, 'It is her death or my own that is being announced to me!' I don't know for what reason, but the idea struck me that the second thought was true, and that this was meant to happen the following day at the same hour.[11]
>
> (Nerval 294)

The idea comes to him that Aurélia might die, and when his premonition becomes reality, he is convinced that these hallucinations are a manifestation of his ill will towards Aurélia, equal parts despair, guilt and thwarted desire. The anxiety that he feels at the loss of Aurélia and his break from reality which result from this trauma have clouded his understanding:

> I don't know how to explain that in my thoughts, earthy events seemed to coincide with events of the supernatural world; these things are easier to sense than to articulate clearly. But just who was this being who was myself and yet outside myself?[12]
>
> (Nerval 316)

Here, one can easily apply the disruptive elements of the unconscious as perceived by Freud in his essay *Das Unheimliche* (*The Uncanny*, 1919), where he lays out a textual analysis of *The Sandman*, to the narrative events of *Aurélia*.

> Uncertainty as to whether an object is animate or inanimate, which we were bound to acknowledge in the case of the doll [Olimpia], is quite irrelevant in the case of this more potent example of the uncanny. It is true that the author initially creates a kind of uncertainty by preventing us – certainly not unintentionally – from guessing whether he is going to take us into the real world or into some fantastic world of his own choosing.
>
> (Freud 138)

Freud insists that Nathanael's unreason, recognized as an objective truth, at least by the narrator, is called into question. Was he the only one to recognize the evil and thus endure suffering at the hands of Coppelius/Coppola? Since Freud's gloss on Hoffmann's tale, interpretations have varied (Ringel 197). For Freud, Olimpia's animation is simply an element of Nathanael's subjective fantasy. Relative to her perceived independence, her romantic feelings for Nathanael are thus also a projection on his part and a symptom of his psychosis.

Ironically, Freud's insistence that the Coppelius/Coppola thread of the narrative is a more compelling example of the Uncanny returns the story to the realm of the fantastical. The psychoanalytical elements of the story, namely Nathanael's projection onto Olimpia, can be explained rationally through hypnosis. By hypnotizing Nathanael, Coppelius/Coppola gains the ability to further manipulate him, preventing Nathanael from going to the police and exposing him as a criminal. However, if Coppelius/Coppola is indeed a supernatural evil, and his double is the manifestation of the unconquered malevolent force from Nathanael's childhood, he cannot simply be a symptom of Nathanael's psychosis. The events of his childhood affected Nathanael's entire family and are thus objective events. Nathanael's sense of persecution, that Coppelius/Coppola has returned to destroy his life, points to the symptom of paranoia as an aspect of his psychosis. Yet, Freud leans into the fantastical elements, agreeing with Hoffmann that the demonic figure *was* persecuting Nathanael, that Coppola and Coppelius are indeed one and the same (Freud 139). This fracturing of the real into the Uncanny is a response to the 'imperfections of real conditions' (Adorno 10) that literature enables, acting as both a reflection of society and a realm beyond it. Freud, in a roundabout way, asks us to look at the psychical state through a fictional character precisely to extract a tangible diagnosis applicable in the real world. In contrast, Nerval's lived psychosis, where the split between ego and shadow was ostensibly never resolved, is treated to a self-psychoanalysis as both a metatext and an autofiction, further blurring the line between reality and fantasy.

Just as Freud interprets *The Sandman* to explore psychoanalytic themes, Jung's examination of Nerval's nuanced portrayal of trauma comes to light in his 1942 and 1945 lectures on *Aurélia*. During the night of Aurélia's death, just after he has seen her double in the street outside his home, the narrator dreams of a dimorphic, angelic-daemonic, androgynous creature who falls from the balcony of an inner courtyard 'tinged with rosy hues and its wings shimmered with countless changing reflections … [It] resembled the Angel of Melancholy by Albrecht Dürer' (Nerval 295). Jung interprets this as a representation of the Self:

> This being, with its thousand colours … represents complete unfolding, because the Self, which is to realize itself, namely, in the abundance of its qualities and its colours. It can only do so when the man, the human being, gives it actions to perform. Without such actions [the] Self cannot unfold, since the Self appears only in what we do.
>
> (Stephenson 55)

Nerval intuits the divine in humanity, thus his own divine and infinite nature. Freud's Uncanny resurfaces in Nerval's imagery of the Dürer angel. Linked to this last encounter with Aurélia, her earthly farewell, the angel is a messenger and a harbinger of the hellish descent that the narrator will soon undergo. Neither male nor female, the angel represents the hermaphroditic, Platonic ideal of humanity and serves as a warning to the narrator to seek happiness within, rather than through a relationship with Aurélia, who has already rejected him.

The esoterism that Jung unravels in his analysis is both a literary conceit and a reflection of the author's interests, personalizing the narrative. Nerval's study of the esoteric and world religions introduced him to Eastern beliefs such as reincarnation and Shinto spiritism – the animation of inanimate objects by spirits known as *kami* – and it would have seemed plausible if not natural to Nerval to recognize a life force in inanimate objects. His underlying mental fragility, spiritual or psychoanalytical, is mirrored in the narrator's crisis, prompted by Aurélia's death, and can be explained in situ and without difficulty outside the text. The teasing of autobiographical information within an allegedly fictional text is not new to Nerval. He had previously woven fact with fiction to great effect in *Voyage to the Orient* (1851) where fictional narrative and travelogue blur the lines between his reality and his inventions. Nerval uses the phantasmagorical to propel his narrator into coming to terms with the sting of romantic rejection by creating his perfect woman.

Aurélia, in this sense, is no more human than Olimpia the automaton. As an idealized creation of the narrator, she acts equally as the embodiment of his psychosis, as does Nathanael's projection of Olimpia. Aurélia has ceased to be 'an ordinary woman of our century' and is elevated to the status of a Laura or a Beatrice (Nerval 292). In other words, she is no longer a person but an ideal. Aurélia is thus stripped of her agency, in death, forced to be loved and admired, if not also ridiculed and despised, by the narrator reeling from his irreconcilable feelings for her beyond death. During his hospitalization, he reproduces the images of Aurélia 'painted in the guise of a divinity' (Stephenson 143) that come to him in his dreams: 'I tried to shape the body of my beloved out of clay; every morning I had to start my work all over again, for the lunatics, jealous of my joy, took delight in smashing the image' (144). The narrator quite literally draws pleasure from shaping his own fabricated likeness of Aurélia in ways not dissimilar to Ovid's Pygmalion. The Enlightenment conceptions of body-as-mechanism are confronted with a Romantic notion of Aurélia's otherworldly powers, capable of inflicting a manner of spiritual suffering on the narrator beyond the grave. The narrator's shaping of Aurélia, both in her literal, physical likeness and in his memory of her after her death, evokes her femininity in a decidedly objectifying way that calls into question her agency. Aurélia the woman dissolves into Aurélia the love-object, an inanimate being which no longer controls how she is portrayed. The narrator's romantic sensibilities – and his almost masochistic longing to rectify the wrongs done to Aurélia while she was alive – are thus a projection onto Aurélia rather than an innate emotional bond between them. The narrator's seeming obsession with (re)creating Aurélia from his memories is an embodiment of his narcissism or megalomania, as Laurence Porter suggests, but more than that, it is a manifestation of the twin fears of loss and loss of control of the Self. The narrator turns to the archetype of the eternal feminine because his objectification of Aurélia is so engrained – by society as by his impulse for ego preservation – that he is unaware of how he portrays her. Just as Nathanael projects his own feelings onto Olimpia, a lifelike non-entity, so too does Nerval's narrator use the memory of Aurélia to fuel an inherent need for control.

Automatic Writing and Mental Illness

The loss of control that the protagonists feel – Hoffmann's Nathanael at the death of his father and again at the discovery of Olimpia's true nature, and Nerval's narrator at the death of Aurélia and his descent into madness – must be acknowledged and dealt with to be overcome. The act of writing *Aurélia* for Nerval was in a sense a coming to terms with his own psychosis, a form of automatic writing prescribed by his psychiatrist, Antoine Émile Blanche.

Automatic writing, also called psychography, is used in both mediumship to channel spirits and as a tool to tap into the unconscious. Used by the Spiritualist movement throughout the nineteenth century, the term 'automatism' or automatic writing described the conscious effort to produce art through tapping into the unconscious. The term was applied to literature, rather than divination or channelling, by André Breton to describe surrealist works. When Breton and Philippe Soupault published *Les Champs magnétiques* (*The Magnetic Fields*) in 1919, they were communing not only with Freudian psychoanalysis and Hippolyte Taine's occult experiments of the 1870s but also with Nerval when they arrived at their conception of the surreal.[13] The Surrealists picked up where the Symbolists had left off. The notion of melding poetry and image to access higher realms, such as in Verlaine's famous quip 'where the uncertain joins the precise' (Lübecker 68) or as Mallarmé put it, poetry as a 'spiritual instrument' (141), coloured the surrealist notion of the medial, which led in turn to mediumship. The Symbolists gave the Surrealists a blueprint from which to draw their ideals of the poetic image and their fascination with allegory and metaphor (Elder 266). In his 1924 Surrealist Manifesto, Breton seems to imply that insanity or possession by spirits could be analogous to the surrealist creative state, which involved passing into a trance-like state to achieve such experiments in automatic writing.[14] Robert Desnos was particularly adept at ascending to a trance-like state when he wrote, sometimes posing a danger to himself and others (Elder 267). Antonin Artaud, in his magnum opus *The Theatre and its Double* (1938), equally returns to the themes of Surrealism via psychoanalysis, attempting through the theatre to externalize internal struggles of the patient's (perceived or actual) condition (80). The invocation of Romantic literary precursors of the nineteenth century, including Nerval, and the experiments with automatic writing in clinical settings by the likes of psychiatrists Jean-Martin Charcot, Esprit Blanche and later Gaston Ferdière (who treated Artaud) lends a certain literary legitimacy to an otherwise marginal and marginalizing practice.

The idea of automatic writing as a contemporary psychiatric tool of care, typically prescribed in the form of journaling, did have its place in Nerval's recovery from his second psychotic break in 1852, and Blanche can be seen as a psychiatric pioneer simply through his innovative and humanist take on mental illness (Clery-Melin and Waintraub 65). His ministering has left us *Aurélia*. Without this care, Nerval may not have had the opportunity or the courage to produce his masterpiece, although his lucidity was not in question (Warren 133). Just as Nerval's retention of his faculties allowed him to record his thoughts even through the lowest points of his illness, so too do his narrator's descriptions of the asylum and his assimilation into its society prove a connection to the real: 'My room is at the far end of a corridor, inhabited on one side by the inmates and on the other the domestic servants' (Nerval 344).[15] He is a singular presence in the corridor: neither inmate nor servant. He has removed himself from the category of 'inmate' or *fou* and instead inhabits a liminal space, just as in his dreams, and his insanity seems not a foregone conclusion, but a question of semantics.

Nerval extends the metaphor of the dream, while adding other figures of speech to delimit the story as a literary work and not simply the scribblings of a psychiatric patient. His narrative can be seen as a 'literary staging of Nervalian psychopoetics' (Wehr 447) where romantic fantasy and Romanticism collide.[16] Nerval juxtaposes normality and madness in the imagistic, dreamlike scenes of teleportation to the banks of the Rhine (299), communion with long-dead relatives (307, 330), descriptions of walks through Paris (337–8) and moonlit promenades on the grounds of the hospital where he is interred (342), evoking New Testament scripture, Plautus and Babylonian myth when he encounters his double. At each instance, this doppelgänger participates in these distressing moments of the narrator's melancholy and anxiety:

> The battle of two souls has planted this mixed seed in a body that itself displays two similar parts reproduced in all the organs of its structure. There is in every man a spectator and an actor, a caller and a responder.[17]
> (Nerval 316)

The narrator feels an urge towards nothingness, recognized in psychoanalytic terms as the death drive, which acts on him like a heavy weight on his soul. At the end of *Aurélia* and after receiving treatment, Nerval ends his story with hope that his suffering is over: 'Nevertheless, I feel happy with the convictions I have attained, and I compare this series of ordeals I have undergone to what the ancients represented as the notion of a descent to hell' (Nerval 354).[18] The narrator finds solace as he has finally transcended his ordeal at the clinic. This was unfortunately not the case for the author: even with psychiatric care, Nerval ended his life in 1855.

A critique of the state of nineteenth-century psychiatry seems appropriate here. The 1838 *loi sur les aliénés* (law on the 'insane') existed in France for close to two centuries without substantial reform until 2011, under pressure from European authorities (Haustgen 49). Although many innovations in the field of psychiatry took hold in France, such as in the work of Charcot, Philippe Pinel, and Esprit and Émile Blanche, many treatments for conditions ranging from hysteria to schizophrenia to epilepsy were dehumanizing and deleterious to patients.[19]

A look across the Channel to understand English society and its solutions shows that dehumanizing practices were no less common here than on the continent. Famously isolationist in his view of psychiatric treatment, Henry Maudsley begins his *Crime and Madness* (1877) with the analogy that visiting a madhouse is like visiting wild beasts in a zoo. This promotes the idea that the 'madman' reacts to the world with greater violence, greater unreason, and is more likely to commit crimes, to ignore moral sense due to a disruption of the faculties of reason. Working in London,

Maudsley's therapies were considered an improvement on the notorious mistreatment during the eighteenth century of patients at Bedlam, but his commentary is no more enlightened, as he provides a pretext for sequestering the 'insane' from society. The mentally ill, deemed incurable, find themselves not only outside the norms of society but outside its sympathy as well (Maudsley 5). Contempt for mental disturbances and the people who suffer from them has existed in Europe at least since the Middle Ages, where the infamous 'pact with the devil' proffered an explanation for any and all eccentricity, incapacity or alienation.[20]

This grappling with the unknown, the unwanted and the Uncanny, before the advent of psychoanalysis and a scientific explanation of the unexplained, was the domain of literature. Thus, Hoffmann and Nerval, in naming the unnamed through the interrogation of their protagonists' psyches, allow for nuance in deciphering human relationships that had previously remained unexplored.

Conclusion

Despite all efforts at containment, the hidden traumas of society tend to surface in politics and national stances, but most effectively in literature. Even when thematically disconnected from war, industrialization and societal upheaval, literature mirrors the human condition (Zilcosky 15). Thus, literature is itself an attempt to externally control the inner turmoil of trauma, of change, and the fears that accompany the loss of purpose or the loss of Self that can come with a shift of identity through the traumatic or the technological.

Hoffmann and Nerval carefully frame the extraordinary as literary elements in their stories to push the boundaries of the real and thus of our interpretation of reality. Intertextuality, allusion, mimesis, all add to the literary quality of the works. Nerval especially takes pains to ground his allusions in a Romantic literary tradition that highlights the influence of myriad erudite elements. The fantastic and the real mix to create a fluidity between dream, idea and hallucination. Hoffmann is content to emphasize the external: ghosts and phantasms, the supernatural, to create the vivid atmosphere of hallucination that is knowingly fictional but somehow also plausible. Nerval taps into the magical state of creativity, choosing to emphasize the internal, the unconscious. Nerval's lucidity during his dream and waking state shows remarkable resolve to portray a dreamworld that could be real, just as Hoffmann devotes his work to the plausibility of supernatural events. Interpreted sometimes as a symptom of mental illness, sometimes as proof of contact with the beyond, but always in a singular and symbolic way, the concept of the doppelgänger, so beloved by the poets of the 'young France of 1830', incorporates the influence of German

Romanticism to create a Gothic interpretation of the everyday that plays with the presuppositions of the fantastical narrative.

Reacting to sociopolitical events as much as to personal demons, Nerval and Hoffmann interrogate the philosophical questions of their time and in some ways predict the future. The Industrial Revolution, which introduced new ways of life to much of the world, completely changed how one's society and environment interact. Weapons of war and scientific and industrial machines created a link to the mechanical that had not previously existed. Even things taken for granted today, such as closed electrical currents, were considered dangerous and frightening at the turn of the nineteenth century (Röttger 246). In *The Sandman*, Olimpia's lifelike yet uncanny movements pose a danger to Nathanael's worldview precisely because of the unknowable aspects of her autonomy. Are her reactions her own, does Nathanael interpret them for his own benefit, or does he completely imagine them as part of his psychosis? Where does his will end and hers begin, if we are to imagine that she has any will to speak of? Likewise, Nerval's narrator acknowledges that his attempt to exert control over Aurélia results in his losing her. His aggrandized memory of her becomes his own creation, more obstinate than the flesh and blood woman. His creation rebels, and the narrator is left to ponder and mourn his loss.

Changing roles for women in less domestic and more public-facing roles, such as Olimpia's musical performance at the party where Nathanael is formally introduced to her, and Aurélia's career as an actress, are positioned within these fictions as detrimental to the male protagonists' worldview, and yet also as symptomatic of the desire to dominate, analogous to the human domination of nature explored in Romanticism. Backlashes against technological innovation prevailed in disparate guises, from the Luddite movement to Spiritism to proponents of magnetism and hypnotism. Nevertheless, the fascination with the hybridity of the organic and the mechanical, once introduced, could not be quelled. The overwhelm experienced by the technological leaps and bounds of the late eighteenth and early nineteenth centuries, coupled with the political upheaval of this era, is reflected precisely in the twin bodies of Gothic and Romantic literature. This link between the rise of the prevalence of mechanization in the ordinary person's daily life and the rise of the treatment of mental illness could be presumed to indicate a loss of cohesion in the fabric of society. As is in evidence in Nerval's biography and in the fictionalization of the psychotic break in Hoffmann's tale, what had been the fear of the supernatural prior to the Industrial Revolution quickly transferred to the fear of madness. However, this presupposition that technology creates insanity ignores the long tradition of both artistic and technological innovation derived from a personal alienation from society at large. Romanticism uncovered and encouraged such innovation but did not invent it. To confuse correlation with causation in the relationship between these instances of technological innovation and the exploration of mental illness and its treatments is to ignore these human impulses.

Notes

1 See Christopher Perry, 'The Jungian Shadow', *The Society for Analytic Psychology*: https://www.thesap.org.uk/articles-on-jungian-psychology-2/about-analysis-and-therapy/the-shadow/.

2 Plato, 'Symposium', *Project Gutenberg*. Translated by E. Jowett. 7 November 2008 [15 January 2013]. E-Book. https://www.gutenberg.org/files/1600/1600-h/1600-h.htm

3 In Scotland, the term 'double-ganger' has been attested in the writings of Walter Scott (Sayers 205).

4 Consider for example Klopstock's interpretation of the German national myth in *Hermanns Schlacht* (*Herman's Defeat*, 1769); Schiller's *Wallenstein* trilogy (1798–99) which revisits the Thirty Years' War; or Goethe's famous *Faust* (Part I, 1808; Part II, 1832) and *Der Erlkönig* (*The Erl-King*, 1782).

5 Translation mine. The original reads: 'Hoffmann a été le grand rénovateur d'un genre qui n'avait pas varié ses formules depuis le moyen âge, même dans *Faust*. Il a épuré le fantastique en le séparant du merveilleux. Selon la belle expression de Barbey d'Aurevilly, il avait obéi à "une voix qui l'appelait au-delà de l'être", mais il a eu soin, au moins dans ses bons jours, de s'arrêter en deçà du surnaturel, à l'extrême bord du possible, sinon du réel'.

6 Translation mine. The original reads: 'Das ist ein böser Mann, der kommt zu den Kindern, wenn sie nicht zu Bett gehen wollen, und wirft ihnen Händevoll Sand in die Augen, dass sie blutig zum Kopf herausspringen, die wirft er dann in den Sack und trägt sie in den Halbmond zur Atzung für seine Kinderchen; die sitzen dort im Nest und haben krumme Schnäbel, wie die Eulen, damit picken sie der unartigen Menschenkindlein Augen auf'.

7 Translation mine. The original reads: 'Der Sandmann hatte mich auf die Bahn des Wunderbaren, Abenteuerlichen gebracht, dass so schon leicht im kindlichen Gemüt sich einnistet'.

8 See Beth Coleman, 'Race as Technology'. *Camera Obscura*, vol. 24 no. 1 (70), 2009, pp. 177–207.
https://doi.org/10.1215/02705346-2008-018.

9 Original: 'Du lebloses, verdammtes Automat!'

10 In 1836, Adolphe Loève-Weimars produced a French translation of Hoffmann, the *Contes fantastiques* (*Fantasy Tales*); a 'new' translation by P. Christian appeared in 1842, published by Flammarion.

11 Translation mine. The original reads: 'Un soir vers minuit je remontais un faubourg où se trouvait ma demeure, lorsque, levant les yeux par hasard, je remarquai le numéro d'une maison éclairé par un réverbère. Ce nombre était celui de mon âge. Aussitôt, en baissant les yeux, je vis devant moi une femme au teint blême, aux yeux caves, qui me semblait avoir les traits d'Aurélia. Je me dis : "C'est sa mort ou la mienne qui m'est annoncée !" Mais je ne sais pourquoi j'en ai resté à la dernière supposition, et je me frappai de cette idée, que ce devrait être le lendemain à la même heure'.

12 Translation mine. The original reads: 'Je ne sais comment m'expliquer que, dans mes idées, les événements terrestres pouvaient coïncider avec ceux du monde surnaturel … Mais quel était donc cet esprit qui était moi et en dehors de moi'.

13 See Elder, R. Bruce, *Dada, Surrealism, and the Cinematic Effect,* Wilfrid Laurier UP, 2021, 287–8.
14 See André Breton, *Manifestoes of Surrealism,* University of Michigan Press, (1924) 1969.
15 Translation mine. The original reads: 'Ma chambre est à l'extrémité d'un corridor habité d'un côté par les fous, et de l'autre par les domestiques'.
16 Translation mine. The original reads: '[Als] literarische Inszenierung der Nervalische Psychopoetik'. See Christian Wehr, 'Phantasmatische Urbilder: Zur Poetik der Halluzination in Gérard de Nervals *Aurélia*'. *Romanische Forschungen*, vol. 114, no. 4, 2002 , pp. 445–473.
17 Translation mine. The original reads: 'Le concours de deux âmes a déposé ce germe mixte dans un corps qui lui-même offre à la vue deux portions similaires reproduites dans tous les organes de sa structure. Il y a en tout homme un spectateur et un acteur, celui qui parle et celui qui répond'.
18 Translation mine. The original reads: 'Toutefois, je me sens heureux des convictions que j'ai acquises, et je compare cette série d'épreuves que j'ai traversées à ce qui, pour les anciens, représentait l'idée d'une descente aux enfers'.
19 See Ada McVean. 'The History of Hysteria' *Office for Science and Society*, McGill University. https://www.mcgill.ca/oss/article/history-quackery/history-hysteria.
20 Kemp notes that the fall of the Roman Empire returned the masses from a scientific, rational view of mental illness to one rooted in demonology and otherworldly explanations. Whether the authorities, such as ecclesiasts and demonologists diagnosing mental illness, truly believed in demonic possession or used it as a ruse to persecute the mentally ill is debatable.

References

Adorno, Theodor. *Aesthetic Theory.* Translated by C. Lehnert, Routledge, 1984.

Artaud, Antonin. *The Theatre and its Double.* Translated by Mary Caroline Richards, Grove Press, Inc., (1938) 1958.

Barine, Arvède. *Poètes et Névrosés: Hoffmann, Quincey, Edgar Poe et G. de Nerval.* Hachette, 1913.

Bloom, Michelle E. "Pygmalionesque Delusions and Illusions of Movement: Animation from Hoffmann to Truffaut." *Comparative Literature*, vol. 52, no. 4, 2000, pp. 291–320.

Breton, André. *Manifestoes of Surrealism.* U Michigan P, (1924) 1969.

Cheng, Lin. "Human Emotions Projected onto Androids: A Manifestation of Internal Crisis—human–android Interactions in 'The Sand-Man' by E.T.A. Hoffmann as Example." *Neohelicon: Acta Comparationis Litterarum Universarum*, vol. 49, no. 2, 2022, pp. 451–63, https://doi.org/10.1007/s11059-022-00669-3.

Clery-Melin, P., and L. Waintraub. "Private Psychiatry in France." *International Journal of Mental Health*, vol. 38, no. 4, 2009, pp. 63–9, https://www.jstor.org/stable/41345303.

Coleman, Beth. "Race as Technology." *Camera Obscura,* vol. 24, no. 1 (70), 2009, pp. 177–207, https://doi.org/10.1215/02705346-2008-018.

El Bahi, Abderrahim. "Le zèle autobiographique dans *Aurélia* de Gérard De Nerval." *Romanica Olomucensia*, vol. 34, no. 2, 2022, pp. 303–22, https://doi.org/10.5507/ro.2022.025.

Elder, R. Bruce. *Dada, Surrealism, and the Cinematic Effect.* Wilfred Laurier UP, 2021.

Freud, Sigmund. *The Uncanny.* Translated by David McLintock, Penguin Books, 2003, https://www.sas.upenn.edu/~cavitch/pdf-library/Freud_Uncanny.pdf. Accessed 23 May 2024.

Haustgen, Thierry. "Les évolutions institutionnelles et théoriques de la psychiatrie française entre 1838 et 1914." *Annales Médico-Psychologiques,* vol. 172, 2014, pp. 45–51.

Hedley, Douglas. "Platonism, Aesthetics, and the Divine." *Platonism at the Origin of Modernity: Studies on Platonism and Early Modern Philosophy*, edited by Douglas Hedley and Sarah Hutton, Springer, 2008, pp. 269–82.

Hoffmann, E. T. A. *Der Sandmann.* Projekt Gutenberg, (1817) 2024, https://www.projekt-gutenberg.org/etahoff/sandmann/sandman1.html. Accessed 23 May 2024.

Kemp, Simon. "Modern Myth and Medieval Madness: Views of Mental Illness in the European Middle Ages." *New Zealand Journal of Psychology,* vol. 14, no. 1, 1985, pp. 1–8.

Lübecker, Nikolaj. *Twenty-First Century Symbolism: Verlaine, Baudelaire, Mallarmé.* Liverpool UP, 2024.

Maudsley, Henry. *Le Crime et la folie.* Germer Baillière, 1877.

McVean, Ada. "The History of Hysteria." *Office of Science and Society*, McGill University, 31 Jul. 2017, https://www.mcgill.ca/oss/article/history-quackery/history-hysteria. Accessed 7 Aug. 2024.

Mizuno, Hiashi. "'Tout vit, tout agit, tout se correspond'. La Folie poétique dans *Aurélia.*" *Revue d'Histoire littéraire de la France*, vol. 110, no. 2, 2010, pp. 333–49.

Nerval, Gérard de. *Les Filles du feu* suivi de *Aurélia.* Gallimard, (1855) 1972.

Perry, Christopher. "The Jungian Shadow." *The Society of Analytical Psychology,* 12 Aug. 2015, https://www.thesap.org.uk/articles-on-jungian-psychology-2/about-analysis-and-therapy/the-shadow/. Accessed 22 July 2024.

Plato. "Symposium." *Project Gutenberg.* Translated by E. Jowett. E-book, 7 Nov. 2008, https://www.gutenberg.org/files/1600/1600-h/1600-h.htm. Accessed 20 Jul. 2024.

Porter, Laurence M. "Mourning and Melancholia in Nerval's *Aurélia.*" *Studies in Romanticism*, vol. 15, no. 2, 1976, pp. 289–306.

Ringel, Stefan, *Realität und Einbildungskraft im Werk E.T.A. Hoffmanns.* Böhlau Verlag, 1997.

Romero, Aurora B. "Pulled to Pieces: Women and (Un)Emotional Labor in E.T.A. Hoffmann's 'The Sandman'." *Monatshefte*, vol. 115, no. 1, 2023, pp. 65–80, https://doi.org/10.3368/m.115.1.65.

Röttger, Kati. "Techno-Logics and Techno-Magics: Phantasmagoria in the Age of Electricity." *Forum Modernes Theater*, vol. 32, no. 2, 2021, pp. 238–53, https://doi.org/10.1353/fmt.2021.0021.

Sayers, William. "A Hiberno-Norse Etymology for English *Fetch*: 'Apparition of a Living Person'." *ANQ: A Quarterly Journal of Short Articles, Notes and Reviews*, vol. 30, no. 4, 2017, pp. 205–9, https://doi.org/10.1080/0895769X.2017.1336073.

Stephenson, Craig E., editor. *On Psychologcial and Visionary Art: Notes from C.G. Jung's Lecture on Gérard de Nerval's "Aurélia."* Princeton UP, 2015.

Warren, Rosanna, "The 'Last Madness' of Gérard de Nerval." *The Georgia Review*, vol. 37, no. 1, 1983, pp. 131–8.

Wehr, Christian. "Phantasmatische Urbilder: Zur Poetik der Halluzination in Gérard de Nerval's *Aurélia.*" *Romanische Forschungen*, vol. 114, no. 4, 2002, pp. 445–73.

Zilcosky, John. *The Language of Trauma: War and Technology in Hoffmann, Freud, and Kafka.* U Toronto P, 2021.

5 Forms of Computation in Hjalmar Söderberg's and Thomas Mann's Decadent Short Stories

Laura Alice Chapot

Introduction: Automation and the AI Revolution

In November 2023, Google reported that it is now spending more on computation than on human employees (Slater). Some commentators have identified this as a watershed moment comparable to other key moments of transition in modes of production. Just as the Industrial Revolution ushered in a wave of automation of manual labour through mechanization, the current AI revolution is seen as marking a new wave of automation in post-industrial societies where skills and activities of the knowledge economy are automated through AI-powered workflows (Azhar and Warren). A year prior to Google's announcement, OpenAI had launched its initial release of ChatGPT – a new generation of machine learning models that use transformer architecture to process extensive datasets and convincingly carry out tasks previously deemed beyond the reach of computation (such as writing creative or critical texts, summarizing and analysing large swathes of information, creating computer programs and producing data visualizations) through an accessible conversational interface. The release of ChatGPT prompted a deluge of discussion over its potential applications and implications. What safeguards and best practices should be developed in response to these generative technologies? What is to be gained and lost, and who stands to gain and lose the most? What new forms of work, collaboration and creativity might be possible, and what are the risks when offloading to these computational technologies fundamentally human skills such as research, creativity and critical thinking? These questions feel particularly new and urgent when core parts of human learning, identity, expression and sociality are at stake. In her critique of *Surveillance Capitalism*, Shoshana Zuboff warns that 'it is no longer enough to automate information flows *about us*; the goal now is to *automate us*' (8, emphasis in original).

Yet this current wave of automation is not as radically new as it may seem and not only a product of recent AI technologies. In many ways, the AI revolution can be seen as a continuation of a particular culture

DOI: 10.4324/9781003545910-6

of automation that grew to prominence during the Industrial Revolution. This chapter proposes to take a literary and media historical approach to illuminate key formal principles characteristic of this culture of automation and compare how these principles play out in computing technologies and literary texts, focusing on formal principles of decadent short stories[1] by Hjalmar Söderberg and Thomas Mann, and on design principles of the Difference Engine – conceived in the 1780s by German engineer and inventor Johann Müller, again in the 1820s by English mathematician and economist Charles Babbage, and finally built, based on Babbage's designs, by Swedish father-son duo Georg and Edvard Scheutz in the 1830s and 1840s (Lindgren 13, 34, 98).

Media studies and cultural histories of technology contextualize technological infrastructure by refocusing attention on how it intersects with situated knowledges and ideologies. These approaches help us understand not only how technological infrastructure is shaped but also how it can become 'an object we think *with* and *through*, not an object we think *about*' (Mullaney 74, emphasis in original). Tara McPherson, for example, traces the lenticular logic – a logic that allows two things to coexist whilst keeping them isolated and apart – at work in both the design of computer operating systems and contemporary discourses on race and the rise of identity politics in the mid to late 1960s. Focusing on formal aspects of design and discourse allows McPherson to sketch out conversational rather than causal relationships – how different aspects of culture respond to one another systemically rather than causally. Just as McPherson traces the lenticular forms connecting computer architectures and political ideologies, this chapter analyses the modular and algorithmic forms which recur across computing technologies and automation ideologies. For a fuller picture, I add to these analyses a comparison with formal features of literary texts.

In literary studies, Caroline Levine's concepts of form and the affordances of form theorize further the connections between ideological, technical and literary contexts. Levine considers forms as 'abstract organizing principles' that can travel whilst also being specific to particular contexts (7). Forms are both a specific, material instantiation as well as general and abstract. Forms therefore provide a way to move between the intangible and the concrete and to relate broader social formations with local instantiations. For example:

> The prison cell cannot do its work without the hard materiality of metal or stone, but it also operates as an iterable way of organizing experience, a model of enclosure that can and does travel across many contexts. It is both a thing and a form.
>
> (10)

In addition, forms afford 'particular constraints and possibilities', and they 'carry their affordances with them as they move across time and space' (Levine 6). In this way, forms can express and perpetuate particular constraints whilst remaining open to alternative possibilities in new contexts.

By combining media history with this concept of form, I analyse how formal features characteristic of current computational design can be traced in the mathematical and design principles of early computing technologies, as well as in the poetics of decadent literary texts, and how these forms relate to a particular culture of automation that came to dominate modes of production during the Industrial Revolution. This culture of automation sets up and polices a tension between human and machine. This tension serves to bolster an ideology of productivity through a culture of management and control – rather than collaboration and commonality – where machines are part of a toolkit of techniques of management that serve to augment and regulate human limitations towards increased productivity. This culture of automation can be traced in the rhetorical presentation of the Difference Engine – the way engineers and promoters conceived of and marketed the Engine as rectifying human limitations. Crucially, it can also be traced in the design of the engines themselves, particularly in the modular and algorithmic principles of the design. Algorithm is understood as a more general term to refer to a rules-based and procedural practice where specific, explicitly formulated, step-by-step instructions are prescribed in order to channel and discipline behaviour towards the achievement of a predetermined aim (Ellenbogen and Langmead 215; Gillespie 27). Modularity is a more specific design philosophy which consists in decomposing overarching procedures into discrete, basic units that are sequentially interdependent whilst also isolated from one another.[2]

In the following sections, the modular and algorithmic forms that constitute this culture of automation are analysed first in the conception and design of the Difference Engine of Babbage and the Scheutzes and then in the formal structures of literary texts by Mann and Söderberg.

In the Difference Engine, the mathematical method of differences allows complex calculations to be decomposed into patterns of sequentially repeated simpler calculations, which became mechanized in the engines through a system of interlocking wheels. In decadent short stories, these modular and algorithmic principles translate into compositional structures where overarching motifs are decomposed into smaller units that transform as they repeat over the course of a story as well as across a body of stories. The ambivalence and irresolution of these stories suggest how these (de)compositional principles could be read both as a creative process of becoming, learning and reinvention, as well as a process of exhaustion and eventual breakdown. Similarly, framing devices create modular story spaces that are both embedded in and isolated from an overarching story space. These diegetic embeddings can provide ways of expressing unspoken aspects of self whilst also serving to keep these aspects isolated and confined. Rather than techniques of control to achieve 'unerring certainty' (Babbage, '"A Letter"') and bolster productivity, the algorithmic modularity of decadent texts instead explores the potential of these forms for ambiguous exploration and irresolution as well as foregrounding their potential for disconnection. Comparing algorithmic and modular forms across computational technologies and literary texts therefore offers contrasting insights into the affordances of these forms, illuminating how they intertwine productivity with exhaustion, and certainty with irresolution and instability.

Forms of Computation and Automation at the Fin de Siècle

In eighteenth- and nineteenth-century Europe, the nature of labour and the modes of organizing production were fundamentally transformed (Lindgren 43–4, 86). The Industrial Revolution marked a shift from small-scale manual labour by individual craftspeople to large-scale production by a mechanized workforce (Babbage *On the Economy* 120). Mechanization – the integration of machines into labour practices – was a primary driver of this shift (Babbage *On the Economy* iii–iv), but there were different conceptions of the role of machinery in automating production, as exemplified in the Luddite disturbances in the early nineteenth century in England. Since the 1760s, textile workers had incorporated machinery in their labour practices and collaboratively refined and innovated the mechanization of their craft (Lindgren, 44, 86; Smith 3–4). However, in March 1811, textile workers in Lancashire began an intense campaign of machine-breaking and destroyed wide-frame stocking machines in several factories. These new stocking machines, while they increased productivity, negatively affected the quality of the product, and the Luddites, as the machine-breakers came to be known, were concerned about the disrepute this would bring to their trade. The Luddites were therefore not protesting the use of machinery as such, but rather a form of mechanization of their craft that prioritized productivity, efficiency and profit over quality and that bolstered particular modes of sociality: individual responsibility and hierarchies of discipline and control rather than communal collaboration (Smith 3–5). In their own words, the Luddites were protesting 'machinery hurtful to commonality' (cited in Thompson 579).

However, this is not generally how Luddites have been remembered. Instead, the Luddite disturbances have gone down in history as a narrative that pits human workers against machines, where workers are in competition with machines and angrily resist inevitable technological change. Yet the ways in which machinery could be hurtful to commonality did not necessarily have to do with the machines themselves, but more with the way in which machinery was implemented – in other words, with the affordances of forms of mechanization. During the Luddite protests, the military had been deployed as an instrument of control to quell the disturbances, and other, perhaps more covert, practices of management and control were gaining ground in factories that further sedimented tensions between human and machine and established a particular mode of automation. New principles of division of labour were being combined with mechanization as an instrument to discipline and control unruly workers towards increased productivity. This culture of automation deconstructed the human workforce into discrete and interchangeable units that could be trained, regulated and disciplined to work together for maximum productivity through organizational as well as mechanical means (Smith 13–14). The feat of automation in the Industrial Revolution was to enact new modes of sociality premised on competition rather than communalism through its modular and algorithmic forms of organization.

The Difference Engine and Automating Computation

The calculating machines developed by Babbage and the Scheutzes provide a unique case study into this culture of automation which, through modular and algorithmic forms of organization, integrates machinery with principles of labour organization towards increased productivity rather than communality. The conception and design of the Difference Engine drew on the idea that human labour could be improved through structures of control, both mechanical and managerial.

The Difference Engine was designed to automate the calculation and printing of numerical tables (Lindgren 13). Numerical tables were, essentially, portable computing aids, usually in the form of a book. They consisted of rows and columns that mapped out relationships between numbers or units. By the nineteenth century, numerical tables had become essential to many key sectors of activity. Astronomers and mathematicians used specialized tables to carry out complex calculations; sailors used tables for navigation; merchants relied on tables to convert between different currencies or metrical systems; and even insurers used tables to estimate life expectancy (Campbell-Kelly et al. 3–4; Lindgren 17–19, 23). Numerical tables were, however, extremely labour-intensive and time-consuming to produce. It could take over ten years for a team of human calculators to produce specialized tables, necessitating large financial investments by governments. Accuracy was also paramount, and errors could be introduced during both calculation and printing stages. In the calculating phase, multiple human computers would be assigned to calculate the same sections of the table, and a 'comparator' would compare the calculations and finalize the correct version (Campbell-Kelly et al. 4). A similar, equally laborious and time-consuming, process of repeated checking and proofreading by different editors aimed to catch any errors after printing (Lindgren 32–3).

Engineers and popularizers of the Difference Engine presented the automation of table production through mechanization as the solution to these issues. The Difference Engine was conceived in direct competition with manual table production and as a corrective to the human fallibilities of the current system of production. The Difference Engine was born out of a dream where one 'could assign laborious and time consuming figure work to some *assistant*, that *never tired, never made an error* and dealt with the numerical calculations for the *higher* degrees as *swiftly* and *certainly* as those for the first degree' (Georg Scheutz cited in Lindgren 138, emphasis in original). These presentations of the Difference Engine can therefore be seen as aligning with a form of automation that pits human against machine and where machines embody the ideal of productivity: they work fast, do not get tired, do not make mistakes and do exactly as they are told. In its mechanization of calculation, the Difference Engine followed the principles used to automate manual labour in manufacturing: it drew on modular and algorithmic forms to break down, organize and regulate the cognitive labour process. In its design, the Engine combined an existing (and relatively ancient) method of calculation – the method of differences – with contemporary ideas about the division of labour.

The method of differences is a mathematical principle that has been used by mathematicians and table producers since at least the fifteenth century to simplify complex calculations. In essence, the method of differences follows modular principles in the way that it decomposes complex calculations into a pattern of simpler calculations such as sequences of additions and subtractions (Lindgren 70–5). In the early 1790s, French mathematician Gaspard de Prony substantially automated the table-making process, not through mechanization, but by implementing the method of differences according to new principles of division of labour from manufacturing. Essentially, de Prony combined algorithm with modularity and created a 'machine made of humans and guided by an algorithm' (Azhar). De Prony, drawing on principles of mass production from manufacturing to regulate the workflow, distributed the calculations of the method of differences across a workforce divided into groups according to specialization (Campbell-Kelly et al. 5).

Babbage was familiar with de Prony's method of table production. In many ways, Babbage simply mechanized de Prony's organizational logic (Campbell-Kelly 5–8). The Difference Engine mechanized specifically the work of the 'unskilled' human computers who were tasked with the computing of the series of simpler operations according to the difference scheme. The Difference Engines (although they sometimes differed slightly in specific mechanical choices, they followed in general the same basic design) were constituted of layers of interlocking wheels. These interlocking wheels acted as discrete calculating units, sequentially carrying out operations such as additions or subtractions. The calculating wheels were set in motion not by steam, as Babbage had first imagined, but by the physical force of human operators cranking a handle (Lindgren 141, 150, 173). The difference scheme could therefore be calculated through the distribution of physical force across the interlocking wheels acting upon one another to induce small modifications – the turning of a wheel – to compute simple calculations. The results of the calculating units were then transferred to the printing unit where the result layer was typeset and stereotyped for printing (Lindgren 59, 77–8, 151).

The design of the Difference Engine therefore involves algorithmic and modular formal principles that organize and discipline the labour process and translate it into physical and mechanical systems of interlocking wheels acting upon one another. The affordances of these forms in this context were to instantiate and regulate a specified step-by-step process in order to mitigate errors and create certainty. Complex processes were decomposed into discrete, basic constituent units, which were then recomposed following particular patterns of organization and regulated through procedural, managerial structures. These formal principles did have the potential to augment production, allowing for large-scale table-making projects to be carried out in record time, but often at the cost of new kinds of tedious and demanding labour and by enacting alienating modes of sociality that conceal and isolate participants in the process from one another. In the following section, I analyse the affordances of these forms of automation in decadent literary texts.

Forms of Computation in Decadent Literature

Decadence at the fin de siècle is often conceived as a counter-cultural artistic movement protesting the despiritualization of modern industrial culture and the normative bourgeois values of orderliness, self-discipline and rational living through themes and motifs of the bizarre, the artificial, the morbid and the perverse (Kafitz 7, 10–15, 21; Koskimies 112; Călinescu 175). Yet if we consider the formal characteristics of decadent texts, these share similarities with the design principles of the Difference Engine. This section investigates how decadent literature responds to modern industrial culture by analysing the algorithmic and modular forms at play in decadence.

Modular and Algorithmic Forms of Decadence

Form is seen as fundamental to decadence (Constable et al. 11; Bauer 69). As Charles Bernheimer notes: '[i]t is not the referential content of the term that conveys its meaning so much as the dynamics of paradox and ambivalence that it sets in motion' (5). Contemporary critics of decadence at the fin de siècle often describe these formal dynamics in terms similar to the method of differences, where an overarching process is decomposed into basic units of operation and sequentially recomposed according to repeated defined procedures. Paul Bourget, for example, describes decadent style as when: 'the book's unity decays to make way for the independence of the page, where the page decays to make way for the independence of the sentence, and the sentence decays to make way for the independence of the word'[3] ('l'unité du livre se décompose pour laisser la place à l'indépendance de la page, où la page se décompose pour laisser la place à l'indépendance de la phrase, et la phrase pour laisser la place à l'indépendance du mot' 15–16). In this description, an initial whole is sequentially decomposed into basic constitutional units through a repeated, formulaic, step-by-step procedure. The exactly repeated phrasing emphasizes the algorithmic nature of the process, whilst the process itself follows modular logics in which overarching wholes are decomposed into discrete, basic units that are interrelated through small serial modifications – only one word is sequentially added and subtracted: the 'book' becomes the 'page', the 'page' becomes the 'sentence' and so on.

Bourget sees these formal principles as characteristic of literary decadence (particularly in the work of Baudelaire), but he also sees them at work in society at large. Indeed, his description of decadent style occurs in the context of his diagnosis of modern society as in a state of decadence where communal sociality is dissolving into a problematic individualism. For Bourget, decadence is 'the state of a society which produces too small a number of individuals suited to works of communal life' ('l'état d'une société qui produit un trop petit nombre d'individus propres aux travaux de la vie commune' 15). Bourget describes society as an organism that is constituted of hierarchies of smaller organisms. The individual is the smallest unit of the social organism, and if society is to function properly, individual units themselves need to function properly. If the individual units become too independent and are not regulated to function at the service of the overarching social organism, society becomes decadent and unable to function productively. This echoes the ideologies of productivity and theories of management and control characteristic of the emerging industrial culture of automation, where individualism, in terms of individual autonomy and creativity, is seen as a fallibility that needs to be regulated and controlled in order to be a productive individual unit in service of the overall structure. Commonality and well-functioning sociality here rely on the regulation of individual units. Bourget sees the formal principles of literary decadence as illustrative and symptomatic of these broader social processes of increasing individualism and deregulation of the social organism.

Bourget's description of decadence, however, is not the final word. In his own diagnosis of decadence in modern culture in *The Case of Wagner* (*Der Fall Wagner*, 1888), Nietzsche rewrites Bourget's description of literary decadence:

> What is the sign of every *literary decadence*? That life no longer dwells in the whole. The word becomes sovereign and leaps out of the sentence, the sentence reaches out and obscures the meaning of the page, the page gains life at the expense of the whole — the whole is no longer a whole.
>
> (Kaufmann 170, emphasis in original)

> (Womit kennzeichnet sich jede *literarische décadence*? Damit, dass das Leben nicht mehr im Ganzen wohnt. Das Wort wird souverain und springt aus dem Satz hinaus, der Satz greift über und verdunkelt den Sinn der Seite, die Seite gewinnt Leben auf Unkosten des Ganzen – das Ganze ist kein Ganzes mehr.)
>
> (Nietzche 21–2, emphasis in original)

Nietzsche follows the principles of modularity and the algorithm of Bourget's description. He uses the same basic units and procedure of decomposition. Even his very rewriting of Bourget's initial formulation can itself be seen as an intertextual algorithm, taking Bourget as an input and sequentially continuing the algorithmic process by introducing his own small serial modifications. Nietzsche, however, also inverts the procedure – starting where Bourget ended, with the 'word' – and the procedure becomes a more dynamic process of decomposition. There is a tension here between decomposition and vitality: even though he is describing the decomposition of the whole, this process is lively and dynamic, with each basic unit actively enacting the process of decomposition. The repetition of 'whole' at the beginning and end of the process also contradicts the process of disintegration by reintroducing a sense of completion, wholeness and balance.

Modular Motifs

Algorithmic and modular forms are therefore at the heart of literary decadence. Whether within texts or across bodies of texts, overarching motifs are decomposed into series of discrete, basic units that are procedurally recomposed and rewritten through a series of small modifications. However, rather than mechanisms of certainty and control, algorithmic and modular forms in decadent texts can instead be read as setting in motion dynamics of paradox and ambivalence and playing out a more ambivalent mode of productivity which is potentially both creative and destructive – as in Nietzsche above. Algorithmic and modular principles are characteristic of Mann's and Söderberg's creative practice. Both deploy practices of serial revision and rewriting across their early short stories. In addition, they use leitmotivic structuring principles where basic motifs are repeated with variations within and across stories. Their early short stories can therefore be read as a modular corpus that algorithmically interacts: stories are both self-contained units as well as fundamentally interrelated as they echo, revise and reformulate one another and allow multiple versions to coexist. In the remaining sections, I will consider the affordances of these modular and algorithmic forms in Mann's and Söderberg's creative practice.

Algorithmic and modular forms in both computational systems and decadent creative practice are seen as potentially generative or creative methods – methods to facilitate the computing of unknown values or exploring a space of possibilities. In decadent poetics, however, algorithmic and modular forms allow for creative explorations of different possibilities but without any definite solution. Mann, for example, conceived of writing as a laborious but valuable process of discovery where he experimented with 'actual or potential selves and actual or potential intellectual attitudes' (Luke ix). His writing, especially his early short stories, functioned as a heuristic method, in the tradition of Nietzsche's philosophy, that could be deployed to continually destabilize preconceived notions and investigate ideas from different angles (Luke x–xi). Mann's early novellas, especially when considered as a whole, can be seen as iteratively exploring different versions and facets of recurring themes. This method of exploration is often realized through leitmotivic techniques – through the repetition and variation of particular motifs. These repetitions and variations of motifs, as in leitmotif technique in music, act as identifiable markers that trace a process of continuity and change. Whilst in computing contexts, modular and algorithmic techniques aim towards a defined goal of certainty, towards finding a solution to a complex calculation, in Mann's practice, instead of a clear goal or solution, it is the process itself that is foregrounded without necessarily any clear resolution. I focus here as an example on three of Mann's early short stories – 'Little Herr Friedemann' ('Der kleine Herr Friedemann', 1896), 'Tonio Kröger' (1903) and 'Death in Venice' ('Der Tod in Venedig', 1912). Each of these stories are both self-contained yet in conversation with one another as they echo and revise one another and play out different versions of possibility. In particular, all three stories end with a moment of epiphany by bodies of water (a river and the sea) that are simultaneously moments of release and liberation as well as moments of breakdown and collapse.

'Tonio Kröger' explores the dilemma of how to reconcile artistic and bourgeois temperaments and find a balance between discipline and creativity. Tonio is struggling to reconcile parts of his identity which are presented to him as contradictions: 'You artists call me a bourgeois, and the bourgeois feel they ought to arrest me' (194) ('Ihr Künstler nennt mich einen Bürger, und die Bürger sind versucht, mich zu verhaften', 317). An industrious and orderly life is at odds with the presumed tempestuous, passionate temperament of a creative artist. And yet Tonio sees both of these aspects within himself: 'I stand between two worlds, I am at home in neither' (194) ('Ich stehe zwischen zwei Welten, bin in keiner daheim', 317). After a journey back to his childhood hometown, Tonio arrives at a resolution to his dilemma. He describes his insights in a letter to his friend Lisaweta in the final pages of the story. He realizes that these parts of himself need not be so sharply separated and demarcated after all. In fact, bringing these two temperaments together offers Tonio particular challenges but also 'extraordinary possibilities' (194) ('außerordentliche Möglichkeiten', 317). It gives Tonio a unique perspective – he is able to see and pay attention to 'the human and the living and the ordinary' (194) ('zum Menschlichen, Lebendigen und Gewöhnlichen', 318) in ways other artists cannot. Tonio is writing his letter to Lisaweta by the sea, and the story ends with a vision, accompanied by the sound of the sea, in which Tonio visualizes his artistic potential and possibilities: 'I gaze into an unborn, unembodied world that demands to be ordered and shaped' (194) ('Ich schaue in eine ungeborene und schemenhafte Welt hinein, die geordnet und gebildet sein will', 318). Tonio concludes that art can be a way of ordering life, of giving shape and bringing life to all aspects of human experience, even 'the delightfully normal, the respectable and mediocre' (194) ('[das] Angenehm-Normale, das Ungeniale und Anständige', 317). However, the resolution Tonio reaches will not last and both 'Death in Venice' and 'Little Herr Friedemann' offer variations on this motif.

Unlike Tonio Kröger and Aschenbach in 'Death in Venice', Herr Friedemann is not an artist but a businessman working for a timber firm. He is, however, from early childhood, marked as an outsider unable to fully develop meaningful social connections (3–6; 87–91). Instead, he finds refuge in a 'peaceful and well ordered' life (6) ('friedlich und geregelt' 91). He teaches himself to appreciate what life has to offer, the pleasures of small things as well as the beauty to be found in art and intellectual pursuits. This peaceful existence is disrupted by his encounter with Frau von Rinnlingen, the wife of the new military commandant. She shows an unexpected interest in him, and at one of her parties, they find themselves together by the river in her garden. For the first time, he realizes that his life of contented and peaceful orderliness was 'a lie and an illusion' (26) ('Lüge und Einbildung', 117). He is seized by a paroxysmic impulse and kneels before her, trying to confess his emotions. She simply 'stare[s] intently straight ahead, beyond him, into the distance' (27) ('blickt[e] starr und gespannt gradeaus, über ihn fort, ins Weite' 118) and then suddenly detaches herself from him, throws him to the ground and leaves 'with a short, proud, scornful laugh' (27) ('mit einem kurzen, stolzen, verächtlichen Lachen', 118). He tries to pick himself up but collapses at the water's edge. Finally, he dies by letting his body fall into the river while, in the background, 'the muted sound of laughter' (27) ('gedämpftes Lachen', 119) emanates from the party.

'Death in Venice' picks up where 'Tonio Kröger' left off whilst also incorporating elements from 'Little Herr Friedemann'. Aschenbach has carried out the resolution Tonio had reached, and lived, similarly to Friedemann, an ordered, disciplined and respectable life which has allowed him to be artistically productive and successful. But it has been at the cost of suppressing aspects of himself, and even his art has now reached a dead end. The story traces a process of both self-realization and decline as he embarks on a journey to Venice and becomes infatuated with a young boy named Tadzio. Aschenbach's story ends, as the title of the novella had announced, with his death. This death by the sea rewrites Friedemann's death by the river. Both endings involve silent exchanges and gestures between the main protagonist and the character who brought about their demise. Aschenbach's death scene, however, is more ecstatic and combines both Friedemann's and Tonio Kröger's scene by the sea, as Aschenbach experiences a hallucination in which he transfigures Tadzio into a vision of Hermes inviting him out towards the sea and its auspicious enormity (267; 157). After this, Aschenbach collapses in his chair on the beach. His vision suggests a sense of open and expansive possibility, as did Tonio's ending. Similarly to Friedemann, however, a tension is introduced between the character's emotional expansiveness and the surrounding world of bourgeois respectability that undercuts the pathos of the scene. Instead of the background laughter of the party, Aschenbach dies not on his own terms but with the announcement that 'later that same day the world was respectfully shocked to receive the news of his death' (267) ('Und noch desselben Tages empfing eine respektvoll erschütterte Welt die Nachricht von seinem Tode', 157). The repeated motif of the endings by water across these stories can therefore be seen as a modular and algorithmic technique that plays out a process of exploration around a guiding question. These stories explore different versions of a central dilemma: the question of how to reconcile discipline and creativity, or bourgeois and artistic temperaments. Even when a resolution seems to be reached, it becomes destabilized when considered in the context of other stories. Rather than reaching a single solution, algorithmic and modular forms in these decadent texts allow for a creative process of exploration but without arriving at any final resolution. In addition, algorithmic and modular forms in these texts develop more ambivalent models of productivity and creativity – Aschenbach's productivity and creative success came at a human cost, whereas Tonio's struggles with creativity ended with a productive insight.

Techniques of serial rewriting are also at the heart of Söderberg's creative process. Söderberg repeatedly edited and revised stories, often introducing very small modifications over different versions of the same story. Söderberg's creative process can therefore be seen to explore the effects of small operations of addition or subtraction. For example, the short story 'The History Teacher' ('Historieläraren') was first published in the newspaper *Dagens Nyheter* on 10 November 1894, and then later as part of the short story collection *Historietter* in 1898. The stories are nearly identical, except for the last sentence. Söderberg removed the last sentence from the 1894 version in his revised 1898 version. Both stories are a portrait of a disillusioned and disaffected history teacher. This teacher has seemingly given up trying to impose any kind of authority in class and instead spends lessons telling anecdotes and trying to entertain the students whilst ending up being the butt of the jokes. Eventually, he has a breakdown in class. In the first version, the class goes silent after the teacher's breakdown:

> His world fell apart, there was nothing he could do but cry, and he cried.
>
> The class suddenly became so strangely quiet.
>
> (Det brast för honom, det var ingenting annat att göra än att gråta, och han grät.
>
> Klassen blev med ens så underligt stilla. Shalin 204)

In the second version, the last sentence is removed, and the story ends with the teacher's breakdown. This simple subtraction changes the overall quality and effect of the story: without this last moment of communion with the students, the focus remains on the teacher and his breakdown, his inability to create connection and his isolation from meaningful modes of sociality. Reading the stories together, knowing that the final sentence has been subtracted, foregrounds and heightens the poignancy of the lack of response from the students to the teacher's open despair. The subtraction of silence makes its absence even more deafening.

Algorithmic and modular forms in Söderberg's writing involve both small modifications as well as repetition. Whereas Mann changed the names of his characters, exploring recurring motifs in new guises, Söderberg instead tends to repeat names, sometimes as signals of recurring motifs he is developing over the course of multiple texts. For example, variations on the themes of passivity, disillusionment and the search for a meaningful life are declined across *Martin Birck's Youth* (Martin Bircks Ungdom,

1901) and *Doctor Glas* (*Doktor Glas*, 1905) through recurring characters. Martin Birck, in *Martin Birck's Youth,* is plagued by his inability to bring himself to decide on a direction in life, whereas the eponymous Doctor Glas does indeed decide to take action, although this too has ambivalent consequences. Birck himself recurs in *Doctor Glas*, he is an old friend of the doctor's (as is the character Markel, who also features both in *Martin Birck's Youth* and in *Doctor Glas*). Throughout the novella, Glas has been wondering whether to kill Pastor Gregorius to save the woman he loves. Moments before Glas makes his final decision and murders the Pastor, he purposefully avoids his indecisive friend Birck:

> Entering Gustaf Adolf's Square I threw a glance into the window of Rydberg's, just in case Markel happened to be sitting there over his absinthe, as he sometimes does. But only Birck sat there in front of a glass of lemonade. A bore; I didn't feel the least inclination to have dinner with him tête à tête.
>
> (119)

> (När jag kom på Gustav Adolfs torg kastade jag en blick in genom Rydbergs fönster i tanke att Markel möjligen kunde sitta där vid sin absint, som han brukar ibland, men där satt bara Birck med ett glas citron. Det är en tungus, honom kände jag ingen lust att äta middag med på tu man hand.)
>
> (121)

Glas explicitly emphasizes that he does not want to see Birck because he is languid and gloomy – the translation 'bore' does not quite capture the range of connotations of the idiomatic expression 'det är en tungus'. In his following diary entry on 22nd August, Glas tells us he has finally made a decision and carried it out: 'It's over. It's done. I've done it' (118) ('Det är färdigt; det är gjort. Jag har gjort det' 120). The structure of the repeated sentences is itself algorithmic and modular. The first two sentences echo one another in structure, with the small modification of the last word. The last iteration of the sentence preserves the word 'done' from the previous sentence, but switches the syntax around, finally putting Glas in the subject position.

Similar repeated reformulations recur throughout the novella to trace different stages of Glas's emotional trajectory. In his entry on 23rd June, Glas repeats twice that 'life passed me by' (34) ('livet drog mig förbi' 27–8). This formulation reflects Glas's sense of passivity and his determinist outlook at the start of the narrative. In his entries on 17th July and 7th September (appearing respectively before and after the murder on 22nd August), Glas repeats twice in each entry a more defiant variation of the initial fatalistic motif, this time addressing life in an active voice (although the punctuation further suggests subtle variations in the motif – the earlier one, before the murder, more defiant with an exclamation mark, the later one more subdued with a full stop): 'Life, I don't understand you!' (73–4) ('Liv, jag förstår dig inte!' 71–2) and 'Life, I do not understand you' (146–7) ('Liv, jag förstår dig inte' 151). This leitmotif is given a final variation in the penultimate entry with the sentence 'Life has passed me by' (150) ('*Mig* gick livet förbi' 155, emphasis added). This might seem to bring the repeated variation of the leitmotif full circle, back to the initial sentence; however, the English translation does not quite reflect the Swedish syntax. Swedish syntax is permutational, and Söderberg has switched the subject and the object of the sentence – in this way, even though the sentence is in passive voice, Glas, as object of the sentence, appears in the subject position. This leitmotivic echoing therefore does not quite go full circle and instead ends in an uncomfortable ambivalence. On the surface, nothing has changed for Glas, he is still alone, his love unrequited, and unable to lead the kind of life he desires. And yet everything has changed because of the unspeakable act of murder he has committed.

Mann's and Söderberg's early short stories can therefore be read as following an algorithmic and modular process of serial rewriting, enacting small changes and revisions from one text to another through processes of addition, subtraction and editing, and where overarching themes are broken down into recurring motifs that transform as they repeat through procedures of repetition with variations. These procedures offer creative possibilities as 'a highly regulated method for disordering the senses of a text' in which 'we cannot predict the results' (Samuels and McGann 36), thereby allowing authors to explore different facets of a character or question. But these processes also suggest the limits of this mode of production as they blur creativity with exhaustion, breakdown and disconnection. These modular and algorithmic processes of (de)composition therefore serve both to give shape to themes and characters whilst also perpetually keeping them unstable and in motion, unable to gain stable footing. This contrasts with the 'unerring certainty' of algorithmic and modular forms in computational contexts.

Modular Story Spaces

Questions of sociality, community and connection are central to both discourses of automation and decadence. As mentioned above, the Luddites were concerned about the modes of sociality that new labour practices afforded. Similarly, decadent texts explore characters' challenges in developing meaningful connections – such as Herr Friedemann, who is resigned to a life of social isolation, or Tonio Kröger, who seeks to resolve his disconnection between two social worlds. Comparing modular forms across computer design and decadent literary texts provides insights into the contrasting affordances of these forms in relation to questions of sociality, connection and identity. In computer design, modularity is an advantageous architectural feature that protects the overall system from failure if one part or unit of the system does not work. In decadent texts, modularity also seems to have beneficial effects in providing protected spaces for aspects of self and identity to be expressed. Yet, these affordances of modularity are ambivalent, as modular forms, whilst facilitating disclosure, also keep these moments of expression isolated and partitioned off. Modular forms therefore allow different parts of the story to coexist, but without having to fully recognize one another, which creates an unresolved unease and tension.

Modular forms in decadent poetics are enacted through framing devices, such as mise en abyme, that insert stories within stories. In this way, story spaces are decomposed into discrete units that are interrelated yet isolated from one another. For example, dreams in 'Death in Venice' and *Doctor Glas* create an embedded story space that allows for certain facets of self to be revealed without disturbing the main course of the story. In 'Death in Venice', it is only in a dream that the true nature of Aschenbach's attraction to Tadzio is revealed (259–61; 146–49). After his dream, Aschenbach starts outwardly modifying his appearance and embellishing himself (261–2; 149–50). In an inverse parallel to the quarantine taking over the city of Venice to contain the growing cholera epidemic, Aschenbach is slowly loosening the quarantining of parts of himself he has been so carefully policing all his life. Similarly, in *Doctor Glas*, the dreams Glas recounts often challenge the version of himself that he is presenting in his diary entries. For example, even though he maintains that he does not feel any guilt or doubt in his decision after the murder, his dreams suggest another story. In his entry on 25th August, Glas recounts how he dreamed that his friends concertedly ignored and avoided him – when he tried to approach them or speak to them, they signalled to one another and turned away. He tried to shake it off and rationalized that perhaps they simply had not recognized him, '[b]ut at the same time I realized I was deceiving myself, and that they recognized me only too well. Waking, I burst into tears' (134) ('[m]en på samma gång förstod jag att jag bedrog mig själv, och att de mycket väl kände igen mig. Då jag vaknade, brast jag i gråt' 137). Dreams therefore create a recursive interior space within the narrative which both allows for truths to be revealed whilst also acting as strategies of textual confinement that keep these truths protected, confined and isolated.

Social rejection and isolation are part of Glas's dread in his dream. Modular framing devices, where motifs that opened the stories also recur at the close of the stories in new forms, often highlight questions of sociality, particularly the demands of conformity and normativity. In his entry on 20th September, Glas decries the oppressive posing and posturing of bourgeois society. In a sudden outburst, he complains: 'And then, these masks! They all wear masks. Worst of all, it is their chief merit. I shouldn't like to see them without. No, nor show myself without! Not to them!' (148) ('Och så dessa masker! De gå med mask allesammans. Till på köpet är det deras största förtjänst. Jag skulle inte vilja se dem utan. Ja, inte heller själv visa mig utan! Inte för dem!' 153).[4] He feels oppressed by this posing and longs to unburden himself, to reveal himself truly and fully to someone (149; 154). But in the end, Glas simply returns to the old habits with which he started the story, talking about the weather and daily life. After the summer heatwave which opened the story on 12th June, it is autumn now, and he senses that snow will be coming soon (150; 156). Despite the murder he has committed, and the pangs of unease he sometimes betrays, he seems to have resigned himself to a life of bourgeois normalcy and pretence. The colour white in which the final scene is drenched, and the emphasis on snow, capable of covering a dirty world in a blanket of white, suggests Glas is ambivalently resigned to continue wearing his mask despite the costs.

Similarly, in 'Death in Venice', framing devices play out tensions between bourgeois normativity and control on the one hand, and individual freedom and autonomy on the other. The story of Aschenbach's unravelling is framed by references to the bourgeois world of respectability of which he is a part. At the start of the narrative, attention is drawn to Aschenbach's reputation and standing with the mention of the added title of nobility to his name ('von') since his 50th birthday (197; 61). This is echoed in the final sentence where, after Aschenbach's ecstatic vision, we are told that 'later that same day the world was respectfully shocked to receive the news of his death' (267) ('Und noch desselben Tages empfing eine respektvoll erschütterte Welt die Nachricht von seinem Tode', 157). The discrepancy between the previous ecstatic prose and the matter-of-fact style of the last sentence creates a destabilizing tension. Along with the repetition of 'as if' ('als ob') when Aschenbach narrates his dying vision, which highlights the scene as an unreliable construction and subjective vision, this serves to limit and constrain the expansiveness of Aschenbach's dying vision:

> But to him it was as if the pale and lovely soul-summoner out there were smiling to him, beckoning to him; as if he loosed his hand from his hip and pointed outwards, hovering ahead and onwards, into an immensity rich with unutterable expectation. And as so often, he set out to follow him.
>
> (Luke 267)

> (Ihm war aber, als ob der bleiche und liebliche Psychagog dort draußen ihm lächle, ihm winke; als ob er, die Hand aus der Hüfte lösend, hinausdeute, voranschwebe ins Verheißungsvoll-Ungeheure. Und wie so oft machte er sich auf, ihm zu folgen.)
>
> (Reed *Der Tod* 157)

Both in 'Death in Venice' and in *Doctor Glas*, the ambivalent endings create a sense of unease under the surface of bourgeois respectability. Framing devices and switches of perspectives allow contrasting realities to coexist but without letting them become fully integrated. And yet, even though these modular forms isolate and confine elements within the story, the reader does not remain unaffected. The modular forms in decadent poetics do not only isolate and confine and conceal, but also invite the reader 'to take two opposite views simultaneously' (Luke xxxix).

Modular and algorithmic forms in decadent poetics therefore illuminate affordances of these forms not always readily apparent in computational and industrial contexts. The ambivalence of decadent stories holds in tension the creative, expressive, and heuristic possibilities of modular and algorithmic forms with their breakdowns and limitations, especially in terms of fulfilment and connection. In addition, rather than simply acting as structures of management and control, decadent poetics suggest how modular and algorithmic forms enable both control and indeterminacy, both channelling and structuring behaviour whilst also allowing irresolvable tensions to coexist that destabilize any secure footing and common ground. Characters and readers are left unsure which direction to take and which versions to believe. As Jacques Dubois notes: decadents 'knew where they came from, even though they did not know where they were going' ('savent d'où ils viennent faute de savoir où ils vont' 384). The computational forms of decadent texts can therefore be read as instigating movements of thoughts, ideas, emotions without necessarily providing any firm resolution. Whereas algorithmic modularity in cultures of automation might emphasize control, certainty and unambiguity, decadent texts demonstrate how a combination of certainty and uncertainty are at play, which feeds a particular sense of alienation and disorientation.

Conclusion: Machinery Beneficial to Commonality

This chapter considered modular and algorithmic forms in fin de siècle decadent literature and contemporary computational technologies to investigate how these forms relate to a particular culture of automation, which deploys structures of management and control to channel behaviour towards increased productivity. Comparing these forms across ideological, technological and literary contexts illuminates the contrasting possibilities and constraints they enable and suggests other potential affordances. This literary and media history of computing technologies at the turn of century therefore seeks to highlight how specific features of computational design, such as algorithms and modularity, can be considered from a literary and media historical perspective as aesthetic forms that perpetuate particular ideological tenets whilst remaining open to alternative possibilities. The aim is to provide contextualizing perspectives on automation, away from dichotomies between human and machine, and towards understanding the premises of such dichotomies and the structuring principles that inform the foundations of these discourses of automation. This historicization of our current AI moment aims to illuminate both its specificities and its continuities with the past. A common rhetorical strategy is to present these technologies as radically new and unique. But this focus on newness obfuscates the existing resources we have at our disposal to incorporate disruptive technological change into our practices and the possible paths that can be taken (Nguyen). Understanding how these techno-social regimes deploy forms hurtful to commonality can open pathways to imagining machinery beneficial to commonality.

Notes

1 Even though Söderberg's *Doctor Glas* (*Doktor Glas*, 1905) has the subtitle 'en roman' ('a novel'), I consider it here as a novella, or a long short story.
2 Modularity remains a key principle in philosophies of computational design to this day – e.g., it is a part of the UNIX design philosophy. See McPherson.
3 Unless otherwise specified, translations are my own.
4 Tonio Kröger also refers to his face as a mask (171; 48).

References

Azhar, Azeem. "Algorithms and Hairdressers." *Exponential View*, 6 Sep. 2021, https://www.exponentialview.co/p/algorithms-and-hairdressers. Accessed Nov. 2024.

Azhar, Azeem, and Nathan Warren. "Will genAI Cause a Compute Crunch? No." *Exponential View*, 9 Oct. 2024, https://www.exponentialview.co/p/will-genai-cause-a-compute-crunch. Accessed Nov. 2024.

Babbage, Charles. "A Letter to Sir Humphrey Davy." *Works of Babbage*, edited by M. Campbell-Kelly, vol. 2, American University Press, (1822) 1989, pp. 6–14.

Babbage, Charles. *On the Economy of Machinery and Manufactures*, 4th ed., Augustus M. Kelley, (1835) 1963.

Bauer, Roger, "'Fin de Siècle' et 'décadence' Comme Catégories Littéraires: Positions et Propositions." *Neohelicon*, vol. 3, nos. 3–4, 1975, pp. 69–86.

Bernheimer, Charles. *Decadent Subjects: The Idea of Decadence in Art, Literature, Philosophy, and Culture of the Fin de Siècle in Europe*. Edited by T. Jefferson Kline and Naomi Schor, Johns Hopkins UP, 2002.

Bourget, Paul. *Essais de psychologie contemporaine*. Plon, Nourrit, 1899.

Călinescu, Matei. *Five Faces of Modernity: Modernism, Avant-Garde, Decadence, Kitsch, Postmodernism*. 2nd ed., Duke UP, 1987.

Campbell-Kelly, Martin, William Aspray, Nathan Ensmenger, and Jeffrey R. Yost. *Computer. A History of the Information Machine.* 3rd ed., Westview Press, 2014.

Constable, Liz, Matthew Potolsky, and Dennis Denisoff. "Introduction." *Perennial Decay: On the Aesthetics and Politics of Decadence*, edited by Liz Constable, Dennis Denisoff, and Matthew Potolsky, U Pennsylvania P, 1999, pp. 1–34.

Dubois, Jacques. "Entre Necessité et Contingence: Le Roman Decadent." *Mimesis et Semiosis: Littérature et Représentation*, edited by Pierre Hamon and Jean-Paul Leduc-Adine, Nathan, 1992, pp. 379–90.

Ellenbogen, Josh, and Alison Langmead. "Forms of Equivalence: Bertillonnage and the History of Information Management." *Technology and Culture*, vol. 61, no. 1, Jan. 2020, pp. 207–38.

Gillespie, Tarleton. "Algorithm." *Digital Keywords: A Vocabulary of Information Society and Culture*, edited by Benjamin Peters, Princeton UP, 2016, pp. 18–30.

Kafitz, Dieter, editor. *Dekadenz in Deutschland. Beiträge Zur Erforschung Der Romanliteratur Um Die Jahrhundertwende.* Peter Lang, 1987.

Koskimies, Rafael. *Der Nordische Dekadent. Eine Vergleichende Literaturstudie.* Suomalainen Tiedeakatemia, 1968.

Levine, Caroline. *Forms: Whole, Rhythm, Hierarchy, Network.* Princeton UP, 2015.

Lindgren, Michael. *Glory and Failure. The Difference Engines of Johann Müller, Charles Babbage and Georg and Edvard Scheutz.* Translated by Crag G McKay, MIT Press, 1990.

Luke, David. "Introduction." *Death in Venice and Other Stories*, edited by Thomas Mann, translated by David Luke, Vintage Random House, 1998, pp. vii–lii.

Mann, Thomas . *Death in Venice and Other Stories.* Translated by David Luke, Vintage Books, 1998.

Mann, Thomas. *Der Tod in Venedig.* Edited by T. J. Reed, Bristol Classical Press, 1996.

Mann, Thomas. *Frühe Erzählungen 1893–1912.* Edited by T. J Reed and Malte Herwig, S. Fischer Verlag, 2018.

McPherson, Tara. "Why Are the Digital Humanities So White? or Thinking the Histories of Race and Computation." *Debates in the Digital Humanities,* 2012, https://dhdebates.gc.cuny.edu/read/untitled-88c11800-9446-469b-a3be-3fdb36bfbd1e/section/20df8acd-9ab9-4f35-8a5d-e91aa5f4a0ea. Accessed Nov. 2024.

Mullaney, Thomas S. *The Chinese Typewriter: A History.* MIT Press, 2017.

Nguyen, Aiha. "Getting Beyond the AI Existential Crisis." *Data & Society: Points*, 3 Jan. 2024, https://medium.com/datasociety-points/getting-beyond-the-ai-existential-crisis-f00235a90166. Accessed Nov. 2024.

Nietzsche, Friedrich. *The Birth of Tragedy and The Case of Wagner.* Translated by Walter Kaufmann, Vintage Books, 1967.

Nietzsche, Friedrich. *Der Fall Wagner: Ein Musikanten-Problem.* C.G. Naumann Verlag, 1888.

Samuels, Lisa, and Jerome McGann. "Deformance and Interpretation." *New Literary History*, vol. 30, no. 1, Winter 1999, pp. 25–56.

Slater, Derek. "Google Reportedly Now Spending More on Computer Power Than People." *GripRoom*, 24 Nov. 2023, https://www.griproom.com/fun/google-reportedly-now-spending-more-on-computer-power-than-people.

Smith, Russell. "Frankenstein in the Automatic Factory." *Nineteenth-Century Contexts*, vol. 4, no. 3, 2019, pp. 303–19.

Söderberg, Hjalmar. *Doktor Glas. A Novel.* Translated by Paul Britten Austin, Anchor Books, Random House, 2002.
Söderberg, Hjalmar. *Doktor Glas.* Albert Bonniers Förlag, 2011.
Söderberg, Hjalmar. "Historieläraren." *Preludier*, edited by Björn Shalin and Nils O. Sjöstrand, Lind & Co, 2021, pp. 201–4.
Thompson, E. P. *The Making of the English Working Class.* Penguin, (1963) 2013.
Zuboff, Shoshana. *Surveillance Capitalism: The Fight for a Human Future at the New Frontier of Power.* Profile, 2019.

6 Prosthetic Verse

Technology, Embodiment and Disability in French Poetry (1984–2024)

Léon Pradeau

Poetry has a unique way of dealing with the entanglement of bodies and technology. Its contribution happens at the level of language and form, beyond representation or conceptualization. Over the last 40 years, poets have attempted to reshape their writing to account for the growing influence of contemporary (bio)technologies on our comprehension of the body. It is easy for us to admit that our bodies and lives are deeply informed by technology – but this understanding has a history. In this chapter, I focus on three experimental French poets whose writing, and whose focus on form, is deeply informed by the technological advances of the second half of the twentieth century: Dominique Fourcade, Anne Portugal and Liliane Giraudon. They published most of their work with P.O.L, which was – at the time, at least – the most important press in France's experimental poetry scene. This article surveys four decades of their writing, during which they explored various ways of providing a formal, historical and political viewpoint on the entanglements between bodies and technology.

Fourcade, Portugal and Giraudon have read – and read with – one another for decades. They have also participated in collective projects. Yet they contribute different perspectives on technology and embodiment. Fourcade has long been concerned with the entanglement between forms and rhythms of technology as extensions of human life, and with poetry as a record of the juxtaposition and exchanges between the human and the non-human. Portugal has insisted on the commodification of human bodies in everyday life (*Les commodités d'une banquette* focuses on signals and injunctions in public spaces; *définitif bob* imagines a character enmeshed in electrical networks and videogame simulations; *selfies* is comprised of a series of photo-linguistic takes).[1] Giraudon's work has focused on forms of social injustice and survival, most of her oeuvre featuring a cast of rural or peri-urban outcasts in gloomy, suspicious atmospheres. In the last decade, following her successive breast cancers, she has focused on language and biotechnologies as tools through which to reflect on illness and disability.

DOI: 10.4324/9781003545910-7

'Prosthesis' is central to these three poets' work for two reasons. First, the prosthetic limb is a visible and functional manifestation of the interactions between technology and human organisms. Fourcade, Portugal and Giraudon are poets who have always worked to give space and representation to contemporary realities such as the rapid evolution of biotechnologies. But prosthesis is also a metaphor that helps conceptualize their own writing as it relates to what I call 'technologies of language', after Marcel Mauss and Bernard Stiegler. Prosthetic 'verse' is the coming-into-form of these broader, societal, humanistic concerns. The representation of material prosthetics is as essential to these poets as the prosthetic metaphor, the history of which is deeply linked to French literature and philosophy.

In *Civilization and Its Discontents*, Sigmund Freud famously called modern human a 'prosthetic god' – geared up and amplified by technological, yet faulty, 'auxiliary organs' (91). This metaphor has travelled throughout Western culture but is most densely represented in the French and Francophone tradition, from Marcel Mauss and André Leroi-Gourhan to Bernard Stiegler, through Jacques Derrida. Let us give a brief account of this lineage and of the growing importance of the prosthetic metaphor. Mauss' 1934 essay, 'Techniques of the Body', argues that the human body is a sum of practices constructed through 'techniques' that are passed on over generations in each culture. His claim challenges the naturality of the human body: it might not be separable from what surrounds it, both culturally and technologically. Retracing the palaeontologic history of the relationship between body and tool, Leroi-Gourhan sees the latter as 'a literal secretion of the Anthropians' body and brain' (*l'outil comme une véritable sécrétion du corps et du cerveau des Anthropiens*) (132):[2] to him, it is not speech or consciousness that defines a human, but standing position and interaction with tools. Derrida refers directly to Leroi-Gourhan when he extends this logic to writing (*Of Grammatology* 84). He proposes that *différance* in writing is analogous to the way tools and prosthetics are grafted onto the organic: a secondary presence which both replaces and supplements (144). Technology remains a detail in Derrida's theory. Only in *Clang* does the supplement appear explicitly represented by a prosthetic limb. Derrida links embodiment and disability with techniques and forms of writing:

> Every thesis is (hardbinds) a prosthesis; what is offered for reading is offered for reading by means of quotations (necessarily truncated, cuttings, repetitions, suctions, sections, suspensions, sewing, grafting, faking, organs without a proper body, a proper body covered with blows, traversed by lice).
>
> (Derrida, *Clang* 189)

Here, Derrida explores the metaphorical prowess of prosthesis, proposing that writing, just like a wooden leg, operates around cuts and grafts through operations that are both surgical and scriptural. Writing, like prosthesis, like 'lice', is like a parasite grafting itself on an origin that cannot be located. Bernard Stiegler is the first to systematize prosthesis not only as a marker of Derrida's *différance* but also as one of the most important biotechnological advances of our time. In *Technics and Time*, he reconciles Derrida's focus on writing with the 'technical tendency' (I, 2) that Leroi-Gourhan first uncovered:

> The evolution of the 'prosthesis', not itself living, by which the human is nonetheless defined as a living being, constitutes the reality of the human's evolution, as if, with it, the history of life were to continue by means other than life.
>
> (I, 50)

Stiegler's definition is compatible with Derrida's vision of writing as a technology that provides a deferred definition of the human by themselves, their life continuing 'by means other than life'. But it also provides a powerful definition of what it means to live with a prosthesis, a part of us that is 'not itself living', yet without which we would fail to be considered human. The 'prosthetic verse' I explore in this chapter stems from Stiegler's (and through him, Derrida's and Leroi-Gourhan's) definition of prosthesis as a non-human provider of humanity, a technological source or supplement for living.

What are the features of prosthetic verse? In this chapter, I do not attempt to fix or generalize the concept. It stands in a grey area between the prosthetic metaphor of writing as technique or biotechnical operation, and the stylistic, technical ways in which poetry can address embodiment, disability, and modern biotechnologies. The three readings are centred on a group of writers who were obsessed with at least two central aspects of prosthetic verse. First, they wanted to think of writing as a material practice, revealing its nature as deferred, constructed, artificial, as opposed to a naturalist expression of affect or experience. This led to many experiments with language and the self that can be read through Derrida and Stiegler's theories of prosthetic construction of language and subjectivity. Second, many writers of the time were invested in thinking through contemporary alterations of the body and the intersections between machines and living organisms. This led to the start of a clash between a poetics influenced by science fiction and fascinated by technological progress and one that focused on navigating new ethical, political, medical concerns about new techniques of the body.

This article explores the large spectrum between a 'technofetishism' deeply anchored in modernity and an attempt at de-technologizing our relationship to bodies. Prosthesis itself is caught between fantasy and materiality. In *The Prosthetic Impulse*, a collective volume from 2007 which problematizes the prosthetic metaphor in light of recent developments in disability studies, Vivian Sobchack writes: 'There are both an *oppositional tension* and a *dynamic connection* between *the* prosthetic as a tropological figure and *my* prosthetic as a material but also a phenomenologically lived artifact' (Sobchack 18). This article, then, also explores the tension between figure and artefact in recent French poetry. Fourcade, Portugal and Giraudon showcase three modes in which prosthetic verse can be conceptualized and challenged. The conversation between them, which now has its own history, allows us to imagine new ways of facing the omnipresent interactions between technology and embodiment.

Prosthetic Extensions of the Self (Dominique Fourcade)

Among recent French poets, Dominique Fourcade has perhaps ventured closest to a language where things and beings are placed in total horizontality and simultaneity. He has consistently understood language, space and time as 'flattened indefinitely, this is the only present or future I can see for myself. For I must identify this body at all costs, this reality, and reach it' [*Indéfiniment à plat, je ne me vois pas d'autre présent ni d'autre avenir. Car il faut à tout prix que j'identifie ce corps, le réel, et que je l'atteigne*] (Fourcade 91). He has chosen '*tout arrive*', both meaning that everything happens (at the same time) and anything arrives, or reaches us, as a key aspect of his poetics.[3] Critics have often used this formula to frame the density and simultaneity of Fourcade's work; all the while proposing that *everything happens* in and through language, densely intertwining writing and reality.[4]

I focus on how this poetics of flattening, simultaneity and horizontality complicates what does and does not belong to the realm of the human – specifically, how human bodies are changed if, and when, *everything happens* to them. The 'reality' that Fourcade might reach is one where the human body is inseparable from its extensions, both in terms of space (their boundaries being questioned) and in terms of time (is there a past in which these extensions were not present). To explore these questions, I will focus on two texts by Fourcade: *Rose-déclic* (*Click-Rose*) (1984), perhaps Fourcade's most famous book; and 'POST-SCRIPTUM', the opening sequence of *Citizen Do* (2008). Technological extensions are particularly present in both but in very different ways, because of their situation in Fourcade's body of work, but also because the cultural imagination of technology had shifted over these 25 years. In my reading of these texts, I demonstrate the essential role of technology in Fourcade's poetics of embodiment and simultaneity, as well as its evolution through his oeuvre.

Rose-déclic was translated into English by Keith Waldrop as *Click-Rose* in 1996. Its title alone is a hybrid between the most classical symbol of poetry and the contemporary techniques of artistic creation. This originary hybrid is essential: while most critics think of technology as being 'incorporated' in *Everything Happens* – hence, technology being *reunited* with other elements of the word, in a post-Heideggerian paradigm – I believe that technology *enables* this poetics. My reading of *Rose-déclic* takes up this challenge.

The second poem of the collection features a systematic structure of the word 'rose' being followed by noun adjuncts: 'Rose brake / Rose neutron bomb' (9). The repetition of this process over short word sequences creates a serial rhythm which is populated by vocabularies of technology: 'brake', 'neutron bomb', 'dizzy-machine', 'step on the gas', 'ejector seat', among others. The central verse of the poem (spreading across three lines) reads: 'Rose of like and of like and of just as click-rose of comparisons exact dizzy-machine (from this immutable all-comparability rising now into the heart of the real) repetition rose'. Consider the length of this verse and the rhythm you would need if you were to pronounce it in one go; this strain is only accentuated by the succession of comparative structures and adverbs. *Everything happens* through the rose because it channels the energy of rhythm and syntax, but also the deeply mechanical, serial rhythm of a 'dizzy-machine' which both repeats and unsettles.

Bernard Stiegler proposes to read space and time as 'decompositions of speed' – because the human impulse is to achieve ever greater speeds, ever greater mobility, composing and decomposing space as these speeds or mobilities are (re)invented (Stiegler, *Technics and Time,* I 17). In Fourcade's case, the speed of perception and writing allows for the *Everything Happens* to produce its intertwining of bodies and technology in the poem – the speed of any 'click' being relative to an individual's perception of sound and sight.[5] In *Rose-déclic*, rhythm is inherently, tautologically technological. There is repetition in the poem, and a 'repetition rose'; halted rhythms and explosive consonants; 'brakes' and 'bombs'. Technology is a supplement to the poem's rhythm: it redoubles its structure by mechanizing it. Reciprocally, in a world ruled by technical objects and processes, the rhythm of a poem can only be repetitive and halted. Both are true of *Rose-déclic,* a book that is heavily marked by the objects and interactions of modern technology. As Walter Benjamin proposes in '*The Work of Art in the Age of Mechanical Reproduction*', 'The equipment-free aspect of reality here has become the height of artifice; the sight of immediate reality has become an orchid in the land of technology' (233). Fourcade's bodies are 'equipped'; the very image of the 'rose' recalls Benjamin's orchid. What do bodies and roses become 'in the land of technology'?

Fourcade's poetry makes a clear move between two important paradigms in both art history and the history of technology in the twentieth century. The timelapse between *Rose-déclic* and *Citizen Do* can be seen as a transition from a modernist paradigm where human beings are 'supplied with' technology (Benjamin 219) to a postmodern paradigm where they are 'supplemented' by it. In Stiegler's words: 'The prosthesis is not a mere extension of the human body; it is the constitution of this body *qua* "human"' (*Technics and Time*, I, 152–3). Between Benjamin and Stiegler, we move from extension to co-constitution. Modernism views technology as a tool of empowerment or extension; postmodernism invites us to read it as an interface operating from within us, reinforcing us in more indirect ways, or ways that are difficult to separate from what we perceive as our selves. According to Benjamin, as a 'test' for the body, cinema prefigures the interlocking of human and technology: 'What matters in these tests are segmental performances of the individual' (246). Several critics have commented on the presence of cinema in Fourcade's work, including questions of staging and montage.[6] But staying with Benjamin's vocabulary, poems in *Rose-déclic* can all be read as 'segmental performances of the individual'. Indeed, both speaker and subject matter are rarely stable from one page to the other; from line to line, or within a single sentence, there can be rapid switches in register or point of view. In the poem 'XIVb', one of the longest in *Rose-déclic*, the rose is attached to a series of nouns which are heterogenous, yet all material and concrete: 'Corundum rose'; 'Rose catapult'; 'velvet rose'; 'Rose plunger' (53–5). This poem constructs a scene where a feminine figure takes off her clothes, yet consistently brings her symbolic 'rose' back to material or technological references – cannons, cars, windows, making up a list of 'neighboring instruments' that are essential elements of the poem's erotic construction. This entanglement of technology and physicality can be related to the notion of 'technofetishism' explored later in this chapter around the work of Anne Portugal. For now, let us remember that this kind of erotic construction of human and machine is a modernist tenet and that the poems in *Rose-déclic* resonate with the proposals of futurism, including Filippo Tommaso Marinetti's:

> Acceleration of life, whose rhythm nowadays is almost always fast [….] Physical, intellectual, and sentimental tightrope of mankind on the high wire of speed, among contradictory magnetisms. Multiple and simultaneous consciousnesses in the same individual [….] Man multiplied by machine. New mechanical sense. Perfect fusion of instinct with the efficiency of motors and with the mastered forces of Nature.
>
> (37–9)

In *Rose-déclic*, we find the desire, speed and mental associations of modernism, linked to a fascination with technology that can be inscribed in the legacy of futurism. The eighties were Fourcade's most strikingly modernist period; already with *Outrance utterance* (1991), he switches to a different approach to rhythm, with longer sentences and a sense of flux that becomes even more marked in *Citizen Do* – the new form of which produces a very different approach to technology, rhythm and embodiment.

In the long poem 'POST-SCRIPTUM', Fourcade seems to relinquish the image of the 'multiplied man' and the dizzying speed of modernist machines. Instead, he uses the figures of prosthesis and amputation to renegotiate the relationship between text and body, space and time, interior and exterior. The poem begins with a meditation on the hardening and softening of stones, which extends to a generalized geology: 'life phrases things in tender-hardness' [*La vie formule des choses en tendresse-dureté*] (*Citizen Do* 9). This movement is reversible and simultaneous, as the stone seems to be hardening while softening, ageing while returning to a primordial state. On the next page, a similar movement is applied to bodies, and specific ones:

> Amputees, return from Iraq, take surfing lessons. And surfers leave for Iraq. My anxiety in the face of such life only grows, especially since I find their body arousing [….] there is nothing I love as much as the amputees' bodies returning, and I understand the motive of my writing is to transcribe the obsessive proximity of such love.
>
> [*Les amputés, retour d'Irak, prennent des leçons de surf. Et les surfeurs partent pour l'Irak. Mon affolement devant cette vie ne fait que croître, d'autant que je trouve leur corps excitant [….] je n'aime rien tant que le corps des amputées de retour, et je comprends que la raison de mon écriture est de transcrire l'obsédante proximité de cet amour*].
>
> (10)

First, this passage depicts the movement of bodies in war as a constant back-and-forth flow between homeland and battlefront, akin to a surf. Though in French, 'surf' only means surfing, Fourcade's mastery of English allows us to extend the meaning of the word: the oceanic surf becomes a key motion in the poem. Bodies come and go, altered always – as if this double entendre had put them in motion. One consequence of the military surf is that war and bodies are always in 'obsessive proximity'. As we know from life and from Fourcade's incessant juxtaposition of contraries and distant realities, terror and arousal often go hand in hand. Should we then be surprised that this segment also turns the spectacle of surfers and amputees into an object of desire? And what to do with the feeling that there is an objectifying perversion in how Fourcade, from his writing desk in France, represents and fetishizes the amputee, which he surreptitiously moves into the feminine?

Among other critics, James Petterson has highlighted that Fourcade's poems 'are replete with references to current events culled from all available media'. In the same article, he also remarks that during the years in which Fourcade was writing *Citizen Do* and his next book, *manque*, 'the massive rise in the number of soldiers receiving prosthetic limbs was extensively covered in the news' (Petterson 18). Similarly to *Rose-déclic*, *Citizen Do* is a book in which Fourcade positions himself as a recorder of the world, registering the most tragic news juxtaposed with the cultural image of the surfer. In many ways, the collage of terror and pleasure is ready-made by information technologies and the society of spectacle.

Fourcade's constant move from observation and representation to projection complicates the above interpretation. In the next segment of 'POST-SCRIPTUM', he switches to the first person as if *he* had been amputated. 'Had they recognized a brother in me?', he asks, before adding: 'I became able to embrace them, in grueling and hopeless succession, uninterrupted to this day' [*Avaient-elles reconnu en moi un frère ?* [...] *J'ai alors pu commencer de les épouser, dans une succession harassante et sans issue, aujourd'hui encore ininterrompue*] (11). The first person in this poem becomes one with the surf of bodies. But this surf is not simply a natural back and forth. With the amputees, Fourcade declares that all motions have motives which preexist them: soldiers did not become amputees by accident, 'they only went there in order to lose something': screams, limbs, parts of themselves. Like the surf of war, each element of Fourcade's poem raises a troubling question: what if loss and suffering were comparable with desire and love, to be sought out through writing, as through life?

In Fourcade's poem, bodies are injured *before* they go to war, as if there were no whole, proper body that might have preexisted the violence of war and amputation. This is also a matter of politics: if bodies always move 'to lose something', it is because they are conditioned and confined by the warmongering politics of the United States at the turn of the century (a politics Fourcade has criticized on various occasions). To Fourcade, this era of political violence and mass slaughter is deeply linked to advances in technological equipment – which is why the image of the prosthesis, as constant and ubiquitous extensions of the self, is so relevant to him as a symbol of the times. Petterson defines this 'time' Fourcade writes in as 'an era of prosthetics and of technologies built into the hardened bodies of amputees equipped with "blade legs"'. However, he continues: 'For his part or, more precisely, for that of Fourcade's "receptors in the stump's nerve endings", they simply were no longer up to the challenge of the prosthetic technologies of this new era' (Petterson 3). Petterson's conclusion about being 'up for the challenge' echoes mine regarding *Rose-déclic*. Writing is always deferred in its quest for novelty and 'permanent innovation' (Stiegler, *Technics and Time*, III, 191). There is simply no way to be fully *in the surf* of new and advancing biotechnologies. This could be a statement about ageing and acknowledging the failures of one's body, but in Fourcade's case, it goes further, since *Rose-déclic* had already demonstrated a form of belatedness in the face of contemporary technology. Fourcade fails in a familiar way: when we try to embrace a reality that is radically new to us, we often fall back to preconstructed categories that are familiar to us. This is what happens with the amputees, in what seems like a compulsion to eroticize the encounter between biotechnology and the 'extraordinary' feminine body. We fall back to a modernist trend that redoubles the extraordinary status of such bodies, instead of staying with their surf: their bodies become spectacles, oddities or sexual projections.

Fourcade's poetic relationship to technology has changed with time. At the start of the 1980s, *Rose-déclic* is all about speed and mechanical rhythms. The result is a constant back and forth between self and supplement, human and machine. Fourcade's early books are made of irregular rhythms that highlight forms of frenetic movements and repetitions of technology and capital. As the title of his 2008 'POST-SCRIPTUM' indicates, times have changed, and so has the perspective: Fourcade begins to look back. Biotechnologies have taken over the fantasy of the machine. Bodies are swaying or surfing with(in) technology: Fourcade's book is a 'total prosthesis'. The flow of this poem is defined by a *vertigo* of the supplement – we can never track down an origin, a former state of mind or body – which is figured by the recursive, ebb and flow of Fourcade's prose. This is a different kind of prosthetic verse which highlights Fourcade's ability to adapt to new forms and rhythms of writing and technology. He is one of the last to continue the futurist and modernist dream of 'machinizing poetry' ('*machiner la poésie*', Guillaume Apollinaire)[7] and adapting to a flow of constant innovation – at the risk of eroticizing the extraordinary and resorting to an aesthetic that most poets would reject today as fetishizing. The case of Anne Portugal will allow us to explore a more deliberate take on this fetishizing tendency in modernist poetry.

Irony and the Eroticization of Prosthesis (Anne Portugal)

In his contribution to *The Prosthetic Impulse*, Marquard Smith introduces the reader to the concept of 'technofetishism', which might remind us of my remarks on Fourcade. Smith defines it as a slippage, variably subtle, between the preoccupation with technology meshing with human bodies and the transformation of this meshing into a fetish. He notes that 'philosophers of technology are prone to take advantage of metaphorical opportunities that are made available by thinking and writing about technology and about the technologization of being' (Smith and Morra 45–6). In 1995, the French poet Anne Portugal provided a concrete example of the 'opportunities' of prosthesis in a short text titled 'Mes robots ménagers' ['My Domestic Robots']. I draw on this text and on Smith's discourse on technofetishism to explore the problematic links between technology, desire and fetishism.

Portugal had published the first version of 'Mes robots ménagers' in 1992, under the title of *Fichier* [*File*] at the confidential press Chandeigne in 1992. But in 1995, it reappeared in the first issue of the *Revue de littérature générale*, titled 'La mécanique lyrique', directed by Olivier Cadiot and Pierre Alféri. Each of the *Revue*'s large pages contains four poems, one at each angle; additionally, each page features one or several grey inserts, linked to poems by arrows, which contain either additional text or images sampled from construction and technical manuals. In the introductory insert on the first page, Portugal presents her new object as an 'electrical' version of the first. She also describes the poems as 'compact units', and the inserts as 'accessories' – metaphorically orienting the ensemble away from a simple medical file towards the realm of technological construction and serial reproduction (Portugal, 'Mes robots ménagers' 173).

In Portugal's book, a masculine speaker manipulates feminine, numbered objects. The 52 numbers might be a count of weeks in a year; but each poem refers to a new bodily encounter, which suggests that a new object-body is invented for each week. The resulting 'file' is a record of their biomedical condition, but also a record of the doctors' offences. Whereas the title might have led us to expect a science-fiction scenario, for instance, with service robots, we start wondering if this doctor is manipulating human patients, or 'domestic robots', dolls, etc. Portugal does all she can to blur the status of these feminine figures. I choose to refer to them as 'objects' because they are both studied and reified: given presence, yet systematically objectified. The slippage between versions of this text materializes the unstable distinction between the realms of the organic and the technological, between integrity and prosthesis.

The first poem of Mes robots ménagers' showcases an object with a 'wooden leg' linked to an inserted 'installation diagram'. On the diagram, multiple pieces are named and represented in their disjointed state, prior to assembly. Throughout the book, diagrams keep hinting at the constructedness of the objects' bodies – this trend being underlined by the poems' frequent references to bodily alterations. For instance, the sixth poem begins with these lines:

> on the sixth one
> I peel off her skin like an orange
> and I wind her onto my wrists
>
> [*à la sixième*
> *je pèle sa peau comme une orange*
> *et je l'enroule autour de mes poignets*] (175).

This poem is linked to a diagram of a rotating motor arm, which features a 'folding mechanism' and a 'tubular axis' [*dispositive de pliage … axe tubulaire*]. Here, the technical object recalls the twisting and folding gesture the masculine subject performs on the sixth object. The presence of the diagrams lets Portugal deconstruct the mechanisms of objectification, highlighting technical processes that run parallel or corresponding to the poems. This is also visible in the seventeenth poem:

> on the seventeenth one egyptian style
> a wise doctor lays out the schematics let's say of the radiator
> to outline
> the end members
> a certified assistant signals it's time
> securely fixated bust
> for all eternity
>
> [*à la dix-septième c'est comme à l'égyptienne*
> *un médecin savant dispose le plan disons du radiateur*
> *pour profiler*
> *les membres terminaux*
> *un assistant assermenté fait signe que c'est bon*
> *bien fixé le buste*
> *pour l'éternité*] (178).

The diagram connected to this poem represents a cabinet connector, with instructions on how to fixate a bolt inside them: optimal angles, depth, etc. In the poem, the words 'securely fixated' respond to these instructions; except in the poem, the three men are not assembling a bookshelf, but a human bust. This situation recalls paintings of dissection rooms, like Rembrandt's *Dr Tulp*, or, more strikingly, J.H. Hasselhorst's *The Dissection of a Young, Beautiful Woman* […] (1864) – with the latter's loaded eroticization of the (deceased) feminine body that we find, here, with the 'egyptian style' and 'eternity' referring to the woman as mummy. But the overall procedural affect of the poem ('lays out', 'outline', 'signals', etc.) is itself ironized, notably by the casual way in which the 'radiator' is picked out as a reference for the construction of this erotic mummy. These three men's supervision, with their haphazard and ironic gestures, provides a good example of the mood of Portugal's book, where women's bodies are constantly reified by failing, perverse masculine objects.

The fetishism of Portugal's book is deeply related to the inanimate or immobilized state of the objects' bodies. Drawing on Freud's 'Fetishism' essay, Emily Apter remarks that fetishism requires 'inanimate or non-human objects, [and] living part[s] of the body [that are] treated as dead or partial objects substituted for the whole'; and that 'these inanimate, non-human, or partial objects are surinvested to the exclusion of all other targets of desire' (Wright 311–14). In Portugal's poems, the inanimate and non-human are explicitly treated as objects made of parts, not as wholes: in fact, the fetishist gaze of the male doctor is specifically targeted at these parts, and at their very status as partial and inanimate, waiting to be (de) composed by him. In the games of sexual domination constructed in the book, there is no space for feminine agency, except futile and comedic acts of resistance (such as 'sulking' [*fait la gueule*]) (179). On the contrary, many poems exhibit forms of non-consensual domination: 'I'm playing the dictator coup'; 'with full authority'; or various forms of the verbal construction *make someone + verb* (*je fais tailler des crayons … je fais passer le bac*) (182–3). 'Mes robots ménagers' creates an erotic equation between prosthesis, lack of agency and objectification. This equation is common in many forms of sexuality, especially in sado-masochistic practices that involve a specific dynamic of submission where the submissive person is 'dollified': that is, reduced to a status of object, a doll made to be manipulated to fulfil its owner's desires. This is literally what happens in the consultation room of 'Mes robots ménagers'. Sado-masochistic practices draw a fine line between sexual play and political dynamics. Having the same distribution for 52 consecutive 'dolls' runs the risk of constructing a space where sado-masochistic practices are simply another name for a version of heteropatriarchy in which man equals dominant and woman equals doll. This is what happens in the pornographic version of dollification, where the latter is often synonymous with non-consensual domination and sexual abuse.

It would be limiting, however, to accuse Portugal of simply falling into these tropes. To conclude my study of 'Mes robots ménagers', I want to explore two ways in which she complicates the technofetishistic and pornographic aspects of her book. The first concerns the status of these women-patient-objects as neutralized by processes of mediatization – in other words, *objectified* in a sense that can no longer be understood as sexual. The second concerns the deep irony with which Portugal treats the whole apparatus of 'Mes robots ménagers'; which affects both its figurations to desire and its political stance. Re-mediation and irony are both techniques of distancing and re-contextualizing within form and discourse, which are essential to Portugal's 'prosthetic verse'.

Pornography seeks efficiency – it needs bodies like recognizable goods in categorized aisles. In *The Prosthetic Impulse*, Elizabeth Grosz contributes an essay on art, pornography, and the extension of the body into media. She proposes that both art and pornography are 'processes of rematerialization – the transformation from one mode of materiality (flesh) to another (paint, canvas, photographic paper, choreographed movements)'. She then opposes art and pornography, adding that: 'Pornography relies on its formula, its theme, its script, to induce its desired effects through a more or less guaranteed pathway. Its eros resides in its insistence on infinite repetition' (Smith and Morra 197). Grosz's concept of repeated rematerialization is easily applied to 'Mes robots ménagers': 52 times over, a woman's body is rematerialized from a patient to an object of fetishist play; 52 times over, a man uses a doll to fulfil his desires of the moment, as a set of parts to be monitored and manipulated in a sexually explicit context. However, 'Mes robots ménagers' adds another layer to Grosz's 'rematerialization': not only are bodies serialized, but they are also constantly modified and fused with less recognizable, linguistic and technological oddities. Each body seems to be processed by scripts and re-processed by its connection with technical diagrams; the latter take over the brevity of the poems which fail to satisfy any pornographic injunction.

Additionally, this serialized re-materialization is marked by a deep sense of irony, which specifically targets the male subject. If there is any pornography in 'Mes robots ménagers', it might be the shooting of a failed movie where the male dominator repeatedly fails at performing his stereotypical role. He cries in the third poem; talks about his hospitalized mother in the fifth; plays dead in the ninth (*à la troisième je pleure* [...] *ma mère est à l'hôpital* [...] *je fais le mort*) (174–6). In the last poem, syntax itself is affected by irony: the line '*je bande avec un boeuf*' (I'm hard with a bull) is a play on the vulgar expression *bander comme un boeuf* (being hard as a bull) – an impotence easily read as a final failure by this man-doctor-actor to perform his dollifying role (186). On the previous page, a poem explicitly stages this failure as mechanical, featuring some typical metaphors of 'prosthetic verse':

> about the forty-fifth under intravenous infusion super
> automatic Paris-Courbevoie interchange
> every day I demand
> a lower price for her gas
> and free maintenance for my small formats

> (*à la quarante-cinquième en perfusion super*
> *automatique échangeur Paris-Courbevoie*
> *je réclame chaque jour*
> *une diminution du prix de son essence*
> *et l'entretien gratuit de mes petits formats*) (185)

In French, '*essence*' means both fuel and essence. This double entendre seems intentional, considering the density of industrial metaphors in 'Mes robots ménagers'. The image of a human essence made of fuel stands as another example of human beings being metaphorized and commodified as technical objects. Beyond this technological metaphor, I find this poem fascinating because it destabilizes the distribution of roles between subject and object. Who has agency? The 45th object is a car (receiving a gas infusion) and a mechanic (providing maintenance): it is occupying both sides of the agency spectrum. There is little space for the male subject. In this poem, he is the victim of repetition: every day, he has to 'demand' savings or free service (which means it never works), not forgetting that the expression 'my small formats' is easily readable as a metaphor for the size of his private parts. In this poem, as in many others, the fetishistic doctor finds himself in the pathetic position of begging an inanimate object for sexual gratification.

In the end, the mechanical apparatus of 'Mes robots ménagers' seems to constantly backfire on the male doctor – the *domestic robots* resisting his mastery, and everyone laughing at his expense. This is an aspect of Portugal's work that was also picked up by Christian Prigent, who commented on the 'nonchalant irony' of her work and situated her between 'the modernists' overheated heroism and the distanced chill' of formalism in the 60s and 70s (Prigent 156). Indeed, though 'Mes robots ménagers' certainly does not abandon the modernist fascination with technological constructs, it ceaselessly attempts to *deflate* it. The over-serialization of human-object encounters, the over-literalization of the object through collages of technical schematics, and the derailing of pornographic tropes, are only some of the techniques Portugal uses to play with the omnipresence of technology and desire in modernist literature.

Portugal's contribution is perhaps the most formally inventive of *La mécanique lyrique*, because it combines a serialized form of writing with a geometrical structure on each page and with the addition of schematics. In this regard, she echoes the modernist fascination for the entanglement between formal innovation and technological experiment. But she turns this formal complexity around, highlighting the malfunctions of such assemblages. Portugal's poems remind us of Freud's image of a 'prosthetic god' whose prostheses give him 'much trouble' – technology as fascinating yet constantly ironized. As such, Portugal's approach resembles the logic of the 'bachelor machine' (which was prefigured notably by Raymond Roussel, and Duchamp's 'The Bride Stripped Bare [...]'),[8] yet more explicitly ironic: in Portugal's writing, the only fusion between human and machine is a farce in which both sides of the prosthesis are malfunctioning and failing to perform their roles.

Disability and Re-Articulation (Liliane Giraudon)

Fourcade and Portugal's works paint a picture of recent poetry struggling to find ways out of the tension between fetishizing technology and viewing it as a dangerous site of dehumanization. Additionally, the very concept of prosthetic verse runs the risk of only standing as a metaphor and obscuring the material reality of living with prostheses and disability. So, we might ask, can the prosthetic trope be used fruitfully regarding sexuality, disability, and the construction of bodies?

Bodies have always been in the foreground of Liliane Giraudon's writing. In her early works, they appear indirectly, as if caught in the background of a picture, sometimes introduced by vague pronouns: '*she* / walks outside in a linguistically / different body' (Giraudon, *Divagation des chiens* 85). There is political weight to this image of a feminine subject (*she*) having to occupy a 'different' body in public space. But such concerns remained discrete and unnamed in Giraudon's early works. She was focused on thinking through her own understanding of the entanglement between bodies and language, organism and grammar, both poetically and politically – taking after Antonin Artaud, Julia Kristeva and other writers who influenced the start of her career.[9]

For this chapter, I focus on some of her later works, especially those of the last decade. In these more recent books, she has focused on exploring this oblique relationship to her own, especially as it relates to sickness and ageing. In *Madame Himself* (2013), she writes about her multiple breast cancers:

> I myself am progressing through a life that is no longer mine, finding myself enclosed in a body that is no longer mine, which illness (its tools and consequences) has rendered not only alien but inadequate.
>
> The body is no longer an ally, it has become an obstacle, and from it I move forward in my vocabulary, slowly dredging a range of scenes and actions I couldn't reach before [....] Childhood rounds up its grammar.
>
> *(Je progresse moi-même dans une vie qui n'est plus la mienne, me retrouve enfermée dans un corps qui n'est plus le mien, que la maladie (ses outils et ses conséquences) a rendu non seulement étranger mais inadéquat. Le corps n'est plus un allié, il est devenu un obstacle, et c'est*

> *à partir de lui que j'avance dans mon vocabulaire, désensablant avec lenteur une variété de scènes et d'actions jusqu'alors inaccessibles* [....] *L'enfance rameute sa grammaire*).
>
> (Giraudon, *Madame Himself* 15–16)

While 'the body' – which often connotes a collective and distanced view of bodies – provides a certain 'vocabulary', Giraudon has explored a new 'grammar', new 'scenes and actions' in her recent books. Her writing now takes place from within, 'enclosed' in a form which she feels is 'inadequate'. In this passage, like in most of Giraudon's books, any understanding of the body is conflated with an understanding of writing. Thinking about, or with one, is thinking about, or with, the other. In *Madame Himself*, perhaps for the first time in her career, the reader is directly confronted by the various ways in which the experience of ageing and disability has displaced Giraudon's approach to writing, from distanced exteriority to subjective interiority.

This brings us back to Sobchack's proposal in *The Prosthetic Impulse*, that any faithful account of prosthesis must move from 'figure' to 'lived artifact' (Sobchack 18). In her essay, Sobchack argues in favour of a reading of prosthetics that reconciles the prosthetic metaphor and her experience as a prosthetic user. She argues that many writers locate disability and prosthesis '*elsewhere*' – quickly dismissing the material, daily experience of disability, in favour of a 'fascination' that we have seen at times in Fourcade, Portugal, and even some of Giraudon's early works. Giraudon's *Madame Himself*, however, *incorporates* the formal practices that had always been at play in her work. For instance, what she calls 'dredging', 'progressing', 'rounding up' are strategies of survival from the hospital, but thee tactics always come from writing. She experiments with a new 'grammar' and 'vocabulary'. For Giraudon, only writing can help reconstruct and reclaim 'a body that is no longer mine'. In an unpublished poem on her website, Giraudon explicitly bridges the gap between recognizing and inhabiting one's body, prosthesis, and the forms of writing:

> No poetry rather this – *I write prose in prose*
> My back to the linden defeated roses
> The other side is this
> Not dead – saying *I no longer look like myself*
> Trembling for him such
> Perfect beauty with no prosthesis
> Place these two under your tongue
> You're losing some of the meaning
> However real under its light

(*Pas poésie plutôt ça* — j'écris de la prose en prose
Dos au tilleul roses défaite
L'autre côté c'est ça
Pas morte — *disant* je ne me ressemble plus
Tremblante pour lui une
Beauté si parfaite et sans prothèse
Mettez ces deux sous la langue
Tu perds un peu le sens
Réel pourtant dans sa lumière.)[10]

Prosthesis stands at the centre of this passage, between 'losing some of the meaning' and recognizing one's body in the mirror. Similarly, moving between prose and poetry responds to a quest for 'meaning' and 'however real' things might be. In his *Prosthesis*, David Wills presents prosthesis as synonymous with articulation: 'there is one thing that comes up time and time again, namely the question of articulation, that is to say prosthesis' (145). In Giraudon's poem, prosthesis articulates the material functions of the body with the operations of writing: after all, articulation is a requirement for movement, but also for comprehensible speech. Wills concludes, in a way that echoes Giraudon's relationship to biotechnologies such as prosthesis, about 'the body in its prosthetic function, in its artificial and contrived articulations, coming like some metallic spectre to haunt the well-sewn surface of originary flesh' (12). Giraudon's experience of disability is focused on this spectre and on various techniques to fight off this haunting. Each articulation is both embodied and artificial, material and literary. In her recent books, she has showcased various techniques of articulation. Let us look at two examples.

In *Madame Himself* (2013), against the body as 'obstacle', Giraudon revisits the Greek myth of the Amazons and their queen Penthesilea.[11] She does not seek to transpose her suffering into their plane but rather to destabilize the neoliberal standardization of histories and mythologies – which also include her, as a disabled, ageing, discarded woman. In Giraudon's book, Amazons are free black women, and instead of having one of their breasts cut off (a 'filthy' masculinist version of the myth), they are 'strong of breast' (*Les amazones deviennent alors celles qui sont « les femmes fortes en sein » « aux seins robustes* [....] *Je me répète à moi-même que le sein tranché était une invention des Grecs, une saloperie incessamment recyclée*) (11–13). The summoning of Penthesilea is a survival strategy: emancipating oneself from patriarchal myths; remaining 'strong of breast' as a survivor of multiple cancers. She adds: 'I go back to the Black Panthers... [...] Gearing up. I think I can give a voice to another Penthesilea, displacing her body into another' (*Je retourne aux Black Panthers...* [...] *M'équipe. Je pense pouvoir donner voix à une autre*

Penthésilée, déplacer son corps dans un autre corps) (14). Throughout the book, Giraudon builds up an analogic displacement between Penthesilea and herself. Stripping the Amazons of their patriarchal burdens is an act of metaphorical surgery where her own survival is at stake – as if these parts of the myths were tumours in her own breast. This is the first technique of re-articulation: deconstructing a myth, creating bridges between women burdened by millennia of constructed ideologies. If Giraudon succeeds in *Madame Himself*, it might be precisely because 'there is nothing left in this text that could fall off detach or be severed' (*Il n'y a plus rien dans ce texte qui puisse tomber se détacher ou être sectionné*) (23). As if she had carefully 'geared up' for a long survival – and found herself lightened by the painstaking act of writing.

Une femme morte n'écrit pas [*A Dead Woman Doesn't Write*] (2023) is a more formal exploration of this recreation of community. First, the book lays them down in verse, making apparent the juxtaposition of voices and discourses. Second, it includes drawings which had been part of Giraudon's creative practice since the 1990s, yet had never been featured in her full-length books. These drawings are collages. They look like outlines from old architecture manuals that Giraudon pasted on paper and drew shapes around. Most of these shapes are abstract and repetitive, though some suggest flowers or a few human figures wearing hats. Giraudon repopulates a set of Greek columns, Italian-style villas, and art deco apartment buildings, with very intricate hand drawings – suggesting how much time and effort was put into each of these small drawings. This collage work echoes that of the poems, which juxtapose utterances visibly taken from heterogeneous sources, as on the last page: 'earth and sky form a single body / in destruction / five exercises / to strengthen / the inside of your thighs' (*ciel et terre forment un seul corps / en destruction / cinq exercices / pour raffermir / l'intérieur de vos cuisses*) (Giraudon, *Une femme morte n'écrit pas* 74). As a final stunt, Giraudon juxtaposes the resistance of a cosmic body and the neoliberal cult of the perfect body (of which women are the first victims). This ironic gesture is characteristic of the whole enterprise: derail every utterance by juxtaposing it with another; rearticulating heterogeneous voices; creating forms of community and resistance. The first two aspects of this poetics of montage are akin to Portugal's endeavour in 'Mes robots domestiques'; but Giraudon, more than most poets of her generation, purposefully directs her poetics towards the (re)formation of collective struggle and survival.

To conclude with critic Jean-Philippe Cazier, these techniques of writing are techniques of the body: 'the body literally becomes a legible text; interior and exterior are constantly pouring into one another, defining one another' [*le corps devient littéralement un texte lisible; l'intérieur et l'extérieur ne cessent de basculer l'un dans l'autre, de se dire l'un l'autre*]

(Cazier 2023). In a way, this resonates with Haraway's proposal, in her *Cyborg Manifesto*, that 'we' don't have bodies anymore, but 'texts and surfaces' (Haraway 245). But Giraudon writes *from* this state of prosthetic bodies, trying to reconstruct something different – as if her body could be reconstructed from these texts and surfaces. The analogy with Penthesilea, the orchestration of mixed genres and voices and the systematized poetics of collage are used as strategies of re-articulation in the face of a dishevelled societal fabric, of an ageing and disabled body. In her *Prosthetic Impulse* essay, Sobchack reminded her readers that what mattered most in living and writing was to 'get on with things both mundane and extraordinary' (Sobchack 38). Giraudon writes with this essential goal. This does not mean repairing an organic matter or a preexisting bodily, or political, reality – which, like Derrida, she sees as illusory; rather, it means that she is constructing new forms of embodiment and political assemblages. Over the last 20 years, she has viewed writing and drawing as techniques of the body, graphic prosthetics on which to stand.

Conclusion

For the last few decades, in France perhaps more than anywhere, prosthesis has stood as both an object and a metaphor. As an object, it has revealed a range of exchanges and influences between human bodies and technological advances; it has complicated the distinctions between bodies commonly defined as abled or disabled, originary or modified. Hence, the importance of the notion of embodiment, which insists on physicality as a constant process of construction and redefinition. As a metaphor, prosthesis has been chosen by various proponents of postmodernism as a concept that does away with the modernist fascination with technology as distinct from the human body, as a pure tool, controllable exteriority. The three poets of this study have different relationships with modernism and technology; but they all conceptualize writing as an act of construction that always neighbours with technology and its advances. Dominique Fourcade is the closest to the modernist fantasies binding machines and human bodies. However, he is also the one whose formalization and conceptualization of technology has changed the most, between 1984 and 2008 – from the jerky gestures of machinic roses to an ambient, incorporated 'global prosthesis' in the Californian surf. Anne Portugal's stance is also rooted in modernism, but she sees its technologic fascinations as targets for irony and nonchalance. When she constructs fictions of desire and perversion and links them to technology, she practices a prosthetic verse that constantly deconstructs itself. While Fourcade moves out of the prosthetic fantasy, she dismantles it from the inside: prosthetic verse becomes a technical operation. The same can be said of Liliane Giraudon, who has always defined writing as biotechnology. She does not conceptualize prosthesis as an exterior spectacle or construction: she has had to live with alterations of her own body, and of bodies around her, for decades. To her, prosthetic verse is a matter of survival: she uses the techniques and supplements of writing to root through, survive through a world overcrowded by social and biomedical constraints. Giraudon does not depict technological or machinic encounters as much as Portugal and Fourcade; but her work offers new ways to bridge the gap between formalist theories of writing (using prosthesis as metaphor) and disability studies (taking prosthesis literally). Ultimately, prosthetic verse refers to the ideal of reconciling formalist understandings of poetry – whose notions of construction and technique are often abstract and disincarnate – with precise ways of embodying the ever-changing links between physicality and technology. But it also refers to a metaphorical failure in which disability ends up being obscured and in which 'the body' risks being universalized and neutralized, both ethically and politically. Liliane Giraudon suggests a way out of this conundrum, but by doing so, she distances herself from technology by focusing on techniques. Contemporary poets seem to be following in her wake as the faith in technology as liberating and reparative reaches new lows. Prosthesis and its conceptual influence on verse are now part of our collective history, haunting us with its 'metallic specter'.

Notes

1 Anne Portugal, *Les commodités d'une banquette*. Paris: P.O.L, 1985; Anne Portugal, *définitif bob*. Paris: P.O.L, 2002; Anne Portugal, *s&lfies*. Paris: P.O.L, 2023.

2 All translations are mine unless otherwise stated. My translations are followed by the French original.

3 See, for instance, the eponymous text, 'Tout arrive'. Dominique Fourcade, *Est-ce que j'peux placer un mot ?* Paris: P.O.L, 2001, p. 57. This text has been translated into English: Dominique Fourcade, *Everything happens*. Trans. Stacy Doris. Sausalito: Post-Apollo, 2000.

4 While Jean-Claude Pinson uses 'tout arrive' as a concept of composition and improvisation (focusing on how *everything happens*), Jérôme Game chooses to use it as a sign of the dense ramifications of reality and perception (*anything arrives*). See Jean-Claude Pinson, 'Le bruit de planeur que fait Dominique Fourcade en écrivant', *Sentimentale et naïve. Nouveaux essais sur la poésie contemporaine*. Seyssel, Champ Vallon, 2002, 237–256 (251); Jérôme Game, 'Dominique Fourcade, or a Poetics of the Rhyzome', in *Poetic becomings*. Bern: Peter Lang, 2011, pp. 107–150 (118–19).

5 Fourcade's 'click' relates to the acceleration of perception in modernity. See Benjamin's use of the concept of 'shock' in modern lyric, in Walter Benjamin, 'On Some Motifs in Baudelaire', III, in *Illuminations*, pp. 160–174.

6 See, notably, Dominique Fourcade, Hadrien France-Lanord and Caroline Andriot-Saillant, *Improvisations et arrangements*. Paris: P.O.L, 2018. 441–3; and Anne Portugal, 'Comme un camion', *Dominique Fourcade: Lyriques*

Déclics. Eds. Laurent Fourcaut, Laure Michel, and Michel Murat. Paris: Hermann, 2020. 251–266.

7 Guillaume Apollinaire, 'L'Esprit nouveau et les poètes', *Mercure de France*, no. 491, 1918: p. 396. This conference has been partially translated into English in *Symbolist Art Theories: A Critical Anthology*. Ed. Henri Dorra. Berkeley: University of California Press, 1994. 308–14.

8 Marcel Duchamp's 1923 *Le Grand Verre* [*The Large Glass*] was originally titled *La mariée mise à nu par ses célibataires, même* [*The Bride Stripped Bare by Her Bachelors, Even*]. On the history of 'machines célibataires', see Michel Carrouges, *Les Machines célibataires* [*Bachelor Machines*]. Paris: MF, [1954] 2024.

9 For a comprehensive study of the links between body and writing in Giraudon's work, see: Léon Pradeau, 'Une époque invalidée? Corps et écriture chez Liliane Giraudon', *Études Francophones*, vol. 31:1, 2024: 101–16.

10 Liliane Giraudon, 'CHANABAJA. Parodie', https://www.lilianegiraudon.com/chanabaja.

11 Her major reference for the character is Kleist's play, *Penthesilea*. See Heinrich Von Kleist, *Penthesilea* [1808]. Trans. Joel Agee. New York: HarperCollins, 2000.

References

Benjamin, Walter. *Illuminations: Essays and Reflections*. Edited by Hannah Arendt and Translated by Harry Zohn, Schocken Books, 1969.

Cazier, Jean-Philippe. "Liliane Giraudon, l'amour à mort. *Une femme morte n'écrit pas*." *Diacritik*, 29 Jun. 2023, https://diacritik.com/2023/06/29/liliane-giraudon-lamour-a-mort-une-femme-morte-necrit-pas/.

Derrida, Jacques. *Clang*. Translated by Geoffrey Bennington and David Wills, U Minnesota P, 2021.

Derrida, Jacques. *Of Grammatology*. Translated by Gayatri Spivak, Johns Hopkins UP, 1974.

Fourcade, Dominique. *Citizen Do*. P.O.L, 2008.

Fourcade, Dominique. *Click-Rose*. Translated by Keith Waldrop, Sun & Moon Press, 2000.

Fourcade, Dominique. *Improvisations et arrangements*. P.O.L, 2018.

Fourcaut, Laurent, et al., editors. *Dominique Fourcade: Lyriques Déclics*. Hermann, 2020.

Freud, Sigmund. *Civilization and Its Discontents*. Translated by Joan Riviere, Hogarth Press, 1953 [1930].

Giraudon, Liliane. *Divagation des chiens*. P.O.L, 1984.

Giraudon, Liliane. *Le travail de la viande*. P.O.L, 2019.

Giraudon, Liliane. *Madame Himself*. P.O.L, 2013.

Giraudon, Liliane. *Une femme morte n'écrit pas*. Al Dante, 2023.

Haraway, Donna. "A Cyborg Manifesto: Science, Technology, and Socialist-Feminism in the Late Twentieth Century." *Simians, Cyborgs, and Women: The Reinvention of Nature*, Routledge, 1991.

Marinetti, Filippo Tommaso. *Les mots en liberté futuristes*. Edited by Giovanni Lista, L'Age d'homme, 1987.

Petterson, James. "Dominique Fourcade: Recalculations." *Studies in 20th & 21st Century Literature*, vol. 43, no. 1, Mar. 2019: pp. 1–24, https://doi.org/10.4148/2334-4415.2026.

Portugal, Anne. *définitif bob*. P.O.L, 2002.
Portugal, Anne. *Les commodités d'une banquette*. P.O.L, 1985.
Portugal, Anne. "Mes robots ménagers." *La Mécanique lyrique*, edited by Pierre Alféri and Olivier Cadiot, P.O.L, 1995.
Prigent, Christian. *Une Erreur de la nature*. P.O.L, 1996.
Smith, Marquard, and Joanne Morra, editors. *The Prosthetic Impulse: From a Posthuman Present to a Biocultural Future*. MIT Press, 2006.
Sobchack, Vivian. "A Leg to Stand on: Prosthetics, Metaphor, and Materiality." *The Prosthetic Impulse: From a Posthuman Present to a Biocultural Future*, edited by Marquard Smith and Joanne Morra, MIT Press, 2006, pp. 17–42.
Stiegler, Bernard. *Technics and Time, 1: The Fault of Epimetheus*. Translated by Richard Beardsworth and George Collins, Stanford UP, 1998.
Stiegler, Bernard. *Technics and Time, 3: Cinematic Time and the Question of Malaise*. Translated by Stephen Barker, Stanford UP, 2010.
Wills, David. *Prosthesis*. Stanford UP, 1995.
Wright, Elizabeth, editor. *Feminism and Psychoanalysis: A Critical Dictionary*. Wiley-Blackwell, 1992.

7 Postcolonial Agency vs. 'French Automation' in Mounsi's *Territoire d'Outre-Ville*

David Spieser-Landes

Introduction: Mounsi's Two-Generation Theory and Double Automation

Law and order, or *la sécurité* in French, has long been referred to in the French media as one of the highest voter priorities in election years, alongside cost of living/spending power (*pouvoir d'achat*), health care and, more recently, terrorism and environmentalism (Mucchielli 2010). This was certainly true in 2007 during the presidential race between right-wing candidate Nicolas Sarkozy and left-wing Ségolène Royal, and remained so in 2024, with Marine Le Pen accusing incumbent President Macron of not doing enough to curb *insécurité* – and linking it to immigration – and Macron defending his political action to improve *sécurité* ahead of the upcoming 2027 election. This theme is often related to *banlieues*, underprivileged parts of French cities with high concentrations of immigrants; people living in a *banlieue* are often the target of discrimination, either racial/ethnic or purely geographical, since *banlieues* are perceived by employers to be in the periphery and actually *outside* of the city (Astier 2023). In his essay aptly called *Territoire d'outre-ville* – translatable as 'French Outercity Territory', a coinage mimicking the common phrase *Territoires d'outre-mer*, i.e., colonized overseas territories such as New Caledonia which has been under the spotlight lately (Corbet and Surk 2024) – Algerian-born French author and rapper Mounsi implies in the very title that colonization of bodies as well as minds is a phenomenon that may well be happening within France as elsewhere. Furthermore, he touches upon his own decolonization trajectory regarding crime specifically and makes his own connections between Franco-Maghrebi youth (*les jeunes*) and *délinquance*/postcolonial violence. At the risk of reinforcing Le Pen's views linking crime and immigration, Mounsi does interpret second-generation immigrant crime as a form of postcolonial *expression* but, in order to make sense of it, he puts it into perspective with first-generation immigrant silencing, oppression and terrible factory work conditions marked by automation. While he does not fully condone or legitimize

DOI: 10.4324/9781003545910-8

postcolonial violence, Mounsi however aims to make sense of it as inherently and organically related to industrial automation. Although *Territoire d'outre-ville* dates back to 1995 and is thus almost 30 years old, the purpose of this contribution is to explicate in detail Mounsi's 'two-generation' argument, with a focus on the nature and impact of automation in the *Trente Glorieuses* era (the 30 years of economic growth between 1945 and 1975), in order to shed light on contemporary sociocultural dynamics in France.

Turning to Mounsi to achieve this analytical goal is a seemingly odd but ultimately very logical choice. In the online abstract for their 2009 book *Parole de Banlieue: Mounsi*, Jeanne-Marie Clerc and Fabien Cruveiller define Mounsi as a 'chanteur de rap devenu romancier' ('a rap singer who became a novelist', Fabula). While Mounsi, born in 1951, is commonly referred to as a rapper, there are, however, also plenty of reasons that arguably justify why, anecdotally yet significantly, the Wikipedia article in English on Mounsi introduces him merely as a 'novelist and singer' – and not as a *rap* singer per se – thus echoing Dominique Le Boucher's characterization of Mounsi, in 'L'Actualité Littéraire', as an 'auteur-compositeur de chansons' ('author and composer of songs') without, again, these songs being necessarily rap songs. In fact, his most famous song, the 1985 'Seconde Génération', which certainly does contain rap themes such as drug consumption and immigrant minority identity, was put to music by Jean-Noël Charléat, who also worked for Alain Chamfort (note the musical similarities between Mounsi's 'Seconde Génération' and Chamfort's 1979 'Manureva', itself very close to Jean-Jacques Goldman's style of music). It comes as no wonder, then, that Mounsi's 'rap' strikingly evokes the 1980s, like a sort of improbable Jean-Jacques Goldman rap, hence classical rap of sorts. From a musical point of view (rather than thematic), Mounsi's rap may thus be best termed as *proto*-French rap, as opposed to rap per se.

In this chapter, I investigate the crucial paradox which I see lying at the heart of Mounsi's literary and artistic production, that is, a strong tension between total compliance to 'French' mainstream ideology[1] and, at the same time, powerful *resistance* to said ideology through his notion of postcolonial 'cri' ('shout'). To do so, I draw upon Jacques Rancière's philosophical apparatus which, in the realm of political aesthetics (the conditions of possibility of what can be perceived/visible/audible), distinguishes between literary and cultural production that strive towards national unity, homogeneity and harmony *versus* those that, on the contrary, introduce linguistic and discursive disharmony within the nation. Rancière calls the former 'archipolitics' and the latter anti-archipolitics, or 'democracy' (Rancière 12). In Rancierian terms, therefore, I scrutinize the tension between archipolitics and anti-archipolitics in Mounsi's works.

In addition to the first-generation immigrant's destructive experience with factory automation, I highlight the second-generation immigrant's rebellious/hybridizing reaction to the crushing force of French cultural assimilationist homogenization, which I characterize as a *machine à broyer les identités*, an identity-crushing *machine*, thus as another form of automation to push against. I show that Mounsi powerfully considers this second instance of automation as something in which cultural freedom may be found – through *problematized* postcolonial violence as well as yet-to-be-defined *banlieue* art and literature.

First Generation-Endured Automation: From Subalternity to Deferred Expression

Although *Territoire d'Outre-Ville* is a broad essay tackling cultural politics at large, it is no coincidence that Mounsi describes in great detail his father's dreadful experience working in the Renault auto industry as a vulnerable immigrant worker in the post-Second World War industrial era. Though an essay, many passages feature novel-like descriptions of the hard-working conditions of these fathers who came to France to find work (cf. the French collocation *trouver du travail*) and, instead of work (*travail*), found extremely harsh work ('emplois pénibles'). In a manner reminiscent of Chaplin's *Modern Times*, but crudely realistic, Mounsi writes, in the autobiographical mode:

> This morning, I saw, on a construction site, two Arab workers [....] I thought of my father. For years, he had worked in one of these factories where a greasy assembly line keeps clinking and rattling without stopping over the workers' heads. Keeping up with the rhythm of the machine, the men's fingers adjust metal parts. They would work in this environment twelve hours every day [...,] with every one of their gestures very precisely pre-calculated, measured and timed. At the end of the day, he would come home – empty. Then every morning, he would leave for work again. One day, the factory shut down. He was fired. But the assembly line continued, for a long time, to go round and round in his worn-out head. It seems that part of himself has stayed caught in that squeaky wheel.
>
> (43–4, all translations mine)

While the sentence 'Then every morning, he would leave for work again' reminds the reader of an automaton, the reference to the consequently worn-out body and 'worn-out head' is reminiscent of Faïza Guène's depiction of her father's premature dementia, for the same reasons, in *Du rêve pour les oufs* (27), showing that this is not an isolated case but concerns, on the contrary, an entire generation of immigrants – the generation of the fathers. Significantly, as it contributes to turning the essay into a poetic/novel-like piece of writing, the assembly line is rendered through a snake metaphor: 'at the end of the long snake' ('au bout du long serpent', 53) the newly produced car finally appears, ready to be driven by the next consumer. Although positively viewed by Renault executives and Renault customers (i.e., a large portion of the French population), the snake is, however, for (immigrant) Renault workers, a harmful creature – even potentially lethal. Renault being an epitome of Frenchness, the narrator's father is thus represented as an exploited victim of a *French* system of super-productivist modernity, culminating in the pathetic description 'my speechless father, tied to his fate, diminished, out of breath, forever crushed under his poor people's burden' (44). In this reality novel, as it were, it is the same real character, Mounsi's father, who, in addition to being reduced to the status (anti-status) of 'subaltern' in the industrial plant, is also shown as 'subaltern' (quite literally as a 'wretched *of the earth*')[2] toiling away in the other context of the coal mine, thus being a character of many lives – or rather anti-lives. Mounsi thus recounts that at the age of 17, in the North of France, his father would go 'down into the mine as a diver holding his breath' – a comparison which underscores the job's danger (44) – and in fact, immigrants 'from everywhere, from Eastern Europe to Northern Africa' are dramatically defined here as 'the coal-trimmers of the industrial revolution' who 'experienced the agony of the depths of the earth well before experiencing that of the tomb' (45).

After linking work for immigrants in the industrial age to nothing less than *death* ('the tomb') and describing the French system of 'integration' as a cannibalistic[3] system of *dévoration*/devouring 'eating up the weak' (58) (as highlighted by the snake metaphor above), Mounsi resorts to one final comparison: Fernand Léger's 1924 film *Ballet Mécanique*. This comparison is significant because this film is part of the Dadaist trend and is commonly known as 'the first movie without a scenario' (Ciné-Club de Caen), thus further underscoring the meaninglessness of the father's life:

> As in Fernand Léger's mechanical ballet, I imagine my father, subaltern [sous-prolétaire] humanoid robot, repeating, with extreme precision, 'machinic' gestures. Having adapted to this automatic mechanism, he himself could not have anything else but the gestures and thoughts of an automaton.
>
> (51)

Their dehumanization through complete economic submission, however, is only one way through which the first generation of immigrants has been disempowered, according to Mounsi. He mentions several other causes.

In a section on France's relationship with Algeria, and more precisely France's tendency to want-to-forget[4] about past colonial crimes and the use of torture, he does not hesitate to use the word *gégène* (31) – a slang term from the French military, short for *génératrice*, meaning 'power generator'. Originally carried on the battlefield during the Algerian War to power portable telecommunication devices, the electricity produced also served as a torture device. Against French forgetfulness about torture (forgetfulness being yet another form of silencing), Mounsi's explicit reference to the *gégène* is itself a shout – a shout which, springing up from the written page, revives the shouts of the tortured fathers. One could argue that the very use of the word 'torture' is a shout in itself, given that Mounsi notes that this very term had been erased from official French language and official French discourse, replaced as it was by the locution 'pression physique modérée' – 'moderate physical pressure' (32).

Mounsi does not enter into *harki* vs. *non-harki* distinctions, thus creating a sort of generic tortured Algerian figure. He notes that, once in France, dehumanization only continued for *all* Algerians, whether previously tortured or not, through a third method of silencing, that is, systematic discrimination: 'In French society, the immigrant worker, although innocent in general, can sense that he is generally considered a suspect. He accepts this very precise and intense little *death of himself*' (66, emphasis added).

A fourth way, related to the third, through which the generation of the fathers has been ruthlessly silenced and disempowered/de-powered consists in police repression of anti-Algerian War demonstrations; the Paris massacre of 17 October 1961 especially occupies a large space in *Territoire d'outre-ville*. Mounsi identifies these events as a 'ratonnade' (22, 31), thus underscoring the plainly racist aspect of this police action (*raton* – a small rat – being a highly pejorative word, especially used in the 1950–1960s, to designate Maghrebi people in French slang). Several dozen demonstrators' corpses (no official figure being available) were thrown into the River Seine as a result of the repression. Mounsi responds to such silencing of the fathers' generation with an account of the *action* (or 'shout') performed by the new generation some 30 years after the silencing occurred:

> On October 17, 1992, we gathered [along the Seine] in Paris. The numerous young people who were present dropped red carnations into Canal Saint-Martin. The drowned dead bodies suddenly swam upstream like salmon .... Young people of immigrant descent have a memory of the Algerian War – a disorderly memory. *The revolt of today's youth is superimposed to something of yesterday*; rebellions add up to each other.
>
> (32–3, emphasis added)

The red colour of the flowers thrown into the water by the sons in remembrance of the father-victims recalls the bloodshed. Their remembering act is literarily depicted as a powerful act – a life-bringing act. Even though the flowers of course will not resuscitate the fathers and do not change themselves into the fathers, they do magically, in Mounsi's imagination and literary production, transform themselves into lively fish, as the phrase 'suddenly swam upstream like salmon' indicates. This comparison is interesting, I suggest, because fish, in French popular language (as in *être muets comme des carpes*, which means 'to be completely silent'), are animals which famously *do not talk*, which means that at this point, the fathers are still, crucially, speech-deprived.

La parole immigrée *Made Visible/Audible: Second-Generation Forms of Expressions*

Describing both the postcolonial situation at large and his own situation as a postcolonial writer, Mounsi takes up again, in a very striking manner, the notion of 'shout', developing it to mean, more precisely, the-shout-of-the-young-immigrant-generation. Importantly, he pairs this 'shout' with what could be termed in contrast the (forced)-silence-of-the-older-immigrant-generation:

> Immigration ties younger and older generations of immigrants together through historical 'interlinkages.' I situate myself within this interstice of time. Our fathers did not shout, did not seek confrontation. They underwent domination in silence. It is this silence which you are urged to decipher in the loud scream [*hurlement*] of their children.
>
> (17)

Postcolonial violence/*délinquance* as a proto-mode of expression is an important concept which bears connections with topics also developed by subsequent rapper Abd al Malik, as I have demonstrated elsewhere (Spieser-Landes 2015). Mounsi's self-avowed criminal record echoes Abd al Malik's equally self-avowed early career as a wallet thief before they both became singers and writers. In Abd al Malik's case, redemption came through his conversion to interpersonal Sufism, while for Mounsi, it is his two-generation theory that helps make sense of it:

> My delinquency as a teenager cannot be understood without reference to the history from which I come [....] Each and every crime I committed was a creative act in relation to the freedom which had been denied to my father.
>
> (65)

This idea is so central that it is rephrased several times: 'I was the other half of my father [...] – a part of himself, and the wild [*sauvage*] part of himself' (2015); 'I only conceived of delinquency as a means to give back to my father some kind of *power*' (67, emphasis added).

To the subalternity and lack of true agency of the fathers, Mounsi responds by highlighting the expression of the second generation – a first level of expression thus being *délinquance* or postcolonial violence.[5] He underscores the sharp contrast between the fathers' lack of being – their subalternity – and, on the contrary, the sense of 'being' which delinquency provides to the second generation: 'Being guilty is, at least, a form of "being". The accusatory finger pointed at you signifies, at least, your singularity' (66).

Evidently, the above assertion runs the risk of essentialism. Mounsi's point, though, is definitely not to imply that all second-generation immigrants *will be* delinquents. However, he certainly affirms that when second-generation immigrant delinquency does occur (as it did for him), it needs to be put in perspective – i.e., explained (hence the two-generation model). Whether explained means *justified* is another debate, but it remains that it is entirely correct to say that, for Mounsi, second-generation immigrant delinquency calls for a sort of collective, diachronic, transgenerational soul-searching. In *Territoire d'outre-ville*, Mounsi expounds the process of such diachronic soul-searching for him as an individual, and subsequently as a writer (thus conceptually passing from subalternity to expression and even, later, literary expression). Note the 'Our fathers/Now, with their sons and daughters born on French soil' opposition in the following quotation:

> French society, it is true, has been used, for centuries, to denying any form of speech to immigrants. It has defined the immigrant as speechless obeyance [*obéissance muette*] and *automatic* submission; a man devoid of any capacity to criticize, judge or reject – an imperfect being, law-abiding not out of good conscience but out of servility. *Our fathers* were naturally seen as analphabets, so their silence was just normal. *Now, with their sons and daughters born on French soil*, French society is challenged to hear voices that do care about the ways in which they are being interpreted.
>
> (17–8, emphases added)

Beyond delinquency, graffiti and *verlan* (a type of slang in which syllables are reversed) – breaking public property and language rules – are presented by Mounsi as further modes of expressions for the second generation, although they are not, according to Mounsi, the end game per se.

Mounsi as Second-Generation Author Both Inhabiting and Resisting the Archipolitics

Verlan, understood as both linguistic and discourse variation relative to the 'purity' of the 'French' language, is seen by Mounsi only as *an intermediate stage* for, in the end, the official normative French language wins, with Mounsi authoritatively saying, using an imperative mode ('must'): 'Someone who owns [a] language consequently owns the world expressed and implied by the vocabulary and syntax [of this language]. It is for this reason that *the youth of the banlieue <u>must</u> appropriate themselves the French language*' (80, emphasis added).

With archipolity being defined, following Rancière in *La Mésentente* (104), as 'unison' or 'harmony' (i.e., monolingual monodiscursivity), *Territoire d'outre-ville* does therefore seem to fully participate in the 'French' archipolitical discourse in which activists in favour of plurilingual polydiscursivity have routinely been described as *machine-like* through the metaphor of the high-prestige idiom (French) acting as a 'rouleau compresseur' (steamroller) to eliminate lower-prestige ones such as regional dialects or Arabic – albeit *arbitrarily* lower-prestige ones (Gaquin 2015). As presented by Mounsi, the reason for this injunction to speak standardized 'French', directed at the immigrant youth of the *banlieue*, is socio-economic in nature – in the sense that speaking standardized 'French' is a necessity if one wishes to find employment and move up the 'French' social ladder (*Le Parisien* 2012).[6] To link this to the regionalist problematic, Mounsi's argument echoes the so-called pragmatic arguments of 'French' discourse in favour of the eventual relinquishing of regional dialects and accents by regional populations (Gaquin 2015). The apparently noble objective is to no longer minoritize both regional populations *and* the immigrant youth, or, as Mounsi would say, make them 'fully French' ('des Français à part entière') instead of being different French people, segregated from the French ('des Français à part', 77). Mounsi encapsulates this thought by linking verlan to a prison of sorts: 'Maintaining the immigrant youth [*les jeunes*] in the use of *verlan*, the language of vagabonds reprimanded by the police, is to condemn them to read the world in an upside-down way' (81).

As a result, to enjoin immigrant youth to join the dominant French language community thus means to *help* (note the positive connotation of the verb) them integrate – integrate into the archipolity. Yet here is the paradox: on the other hand, it turns out that the theme of *accent* – in other words, *anti-archipolitics* – is equally central in Mounsi's literary/artistic production. As much as *verlan* (i.e., a form of accent in the sense, again, of both linguistic and discourse variation relative to the 'purity' of the 'French' language) was vilified as 'non-French' and therefore as an obstacle to integration, it is *at the same time* relativized as an intermediary stage and in fact *extolled* as a sign of great promise for empowered minority expression in its own right; minorities, in other words, do not *need* to disappear any longer:

> The youth of immigrant descent, too, have a culture of their own. And it shall not always be the case that a mysterious Someone [*on ne sait qui*][7] has the power to exclude them or include them. The 'verlan' in which French society [*on*][8] wants to imprison them is only a moment in the history of their language [....] My wish is that they will be able to create a new language, written with so-to-speak stenographic, rhythmic words, extremely fast-paced and at the same time extremely expressive. Like graffiti taggings. They do have the potential, within themselves, to invent this. 'Verlan' is a '*shout*,' a transitory necessity, almost a conditioned reflex. It is a moment in the evolution of oral language. Why wouldn't they exist in the written language? I live in the hope of the written moment of their reality, that is to say the moment when what is nowadays manifested through violent acts will, tomorrow, be able to carry out, through words, *meaning*.
>
> (80, emphases added)

Note, in this central quotation, the conceptual gradation from (1) violence, 'violent acts', as a crude mode of expression but a mode of *proto*-expression nonetheless, to (2) 'oral language' (the first level of expression, verlan being a stage thereof), and to (3) 'written language', considered an ultimate, more elaborate level of expression. As for 'oral language', i.e., *verlan*, Mounsi's characterization of it as 'shout' (cri) gestures towards a shift from standard French language, hence from Frenchness, through the motif of 'jarring dissonance' – or 'accent', in a Rancierian approach. Regarding written language, Mounsi seems to be describing/prophesizing the advent of a kind of postcolonial literature which, in the context of metropolitan France's *banlieues*/urban ghettos, would necessarily be 'slammic' (whether literally or metaphorically) in nature – hence a 'slammic', or 'rap-like' postcolonial literature: *Banlieue* literature (or *Beur* literature, *beur* being the verlan of 'Arab') *as a shout* departing from tamed Frenchness, in the same way that oral language was a 'shout'. *Beur*/*banlieue* literature written by second-generation immigrant writers will be giving a voice – an accented, thus *shouting* voice (in contrast with previous silence) – back to the voiceless fathers.

Proposing here a novel way of reading *Territoire d'outre-ville*, I suggest that the essay, taken as a whole, translates into literature the already mentioned 'diachronic soul-searching' as it not merely talks about the two-generation theory but in effect literarily *performs* it. The very title *Territoire d'outre-ville* is polysemic. A neologism built upon the already existing phrase 'territoires d'outre-mer' (the 'overseas territories' inherited from colonial policies), its first and most obvious meaning is to denote exclusion ('*outre*' meaning 'beyond'), thus denouncing the *banlieue* as an urban space paradoxically *cut* (conceptually and through infrastructures such as beltway roads) from the rest of the urbs/city. However, my reading of the title re-emphasizes that this is not the only way Mounsi understands the phrase.

Indeed, the author resorts to two tropes connected to the idea of stream of consciousness or the unconscious: the labyrinth and the river. The labyrinth ('In the labyrinth of the city, I grope my way through my memories, searching the smallest streets on a quest for my own life', 16–17) is made to mirror the river in their created commonality with the circumvolutions of both Parisian streets and mental streets in which the subject gets lost:

> It took me thirty years to move from the 'Crimes and delinquency' section of the newspaper to the literary section. My memories are forcing me through a windy itinerary to find myself again [*pour me retrouver*] in this part of myself which I call 'Outercity Territory' – for me in particular: 'Territory beyond the Seine River.' And this river continues to flow inside me, shaping me, deeply, secretly.
>
> (21)

It follows that the phrase 'Territoire d'outre-ville' is not purely associated with the physical urban space of the banlieue, but, rather, with an imaginary intermediary space (interface) where the physical, urban space would meet the interior, mental space of the narrator: 'this part of myself which I call "Outercity Territory"', '"Territory beyond the Seine River" [...] flowing inside me', writes Mounsi. In both quotations, the external/internal cityscape is linked to the motif of memories: 'I grope my way through my memories, searching the smallest streets', and 'my memories are forcing me through a windy itinerary'. Last but not least, the city is rendered through a unique metaphor: that of a labyrinth, which can be read as an image for the unconscious (as can be the fluidity of the river flowing inside and shaping the individual). The soul-searching mentioned earlier thus involves even the unconscious: 'No one can ever say these simple words: "I am who I am"' (16).

There is one powerful way through which such foregrounding of the unconscious may uniquely strengthen Mounsi's two-generation theory. It rests on hypnosis-related notions such as collective memory and past life regression (the idea that memories of past lives, or incarnations, may be accessed through hypnosis). My argument is that Mounsi is here using a form of *metaphorical* past life regression theory, as well as a *metaphorical* notion of collective memory, to say that his memories are not just made up of 'his' own memories, but comprise his memories *and* those of his father(s). This interpretation is corroborated by the following excerpt from *Territoire d'outre-ville*'s Foreword: 'From the shores of the Mediterranean to the French Outercity Territories, I patiently attempted to glue back together the pieces of my memory' (11).

Mounsi's memory (archetype of the second-generation immigrant's memory) is recognized as being fragmented and will only be whole again when connected with the memories of his fathers. Such is, in my opinion, the ultimate meaning of Mounsi's key phrase in the essay: 'interlinkages' ('raccords historiques'), in line with recent research – summarized by the BBC asking the question 'Can the legacy of trauma be passed down the generations?' – which highlighted the growing scientific attention put on the role of epigenetics (Henriques 2019) in the study of transgenerational trauma.[9]

Before finding this intergenerational connection through the literary experience, Mounsi confesses that he sought it through drugs (perhaps a hypnosis-like experience, or at any rate a modified state of consciousness), which he thus describes: 'I remember sometimes seeking to forget this obscure and fragmented conscience through using drugs' (14) – an experience which Mounsi readily acknowledges led to a situation of dependence and addiction: 'I would come back from the trip in an ever more terrible state. So, when solitude and anguish were lurking, I would once again "leave my body" and land in a sort of ecstasy' (14). Nonetheless, the author depicts drugs as the locus of a special connection with time, and thus memory:

> I was doing that just in order to vanish, to flow with the sand inside the timer, and dissolve myself out of the present. I however needed to retrieve my lost days in order to fully measure what my history means – the history of which, as the son of an immigrant, I am both an actor and a witness.
>
> (Mounsi, 14).

Related to topics of time and collective memory, the phrase 'pre-history' in what follows is, I argue, especially significant: 'But my pre-history is as difficult to decipher as recently dug-out archeological artifacts' (14).

Now an active writer with a literary agency, as opposed to his former self as a passive drug user, Mounsi then continues to hold the two-generation theory as a major topic in *Territoire d'outre-ville*. This is confirmed in the essay's incipit, when Mounsi describes the writing process as a sort of *recherche du temps perdu* – not only *his* 'lost time', but also his father's lost time, as shown by the notion of 'interlinkages' [raccords historiques'] understood as interstices of time:

> To write is to lay down, for a moment, on a piece of paper, one's internal mess to see if it makes any sense. Time is not a prisoner of clocks or calendars. It runs, or gets lost, within words – some people set off on a journey to find it again [....] Memories long forgotten resurface, entire scenes and complete situations become present.
>
> (13)

Postcolonial literature, in other words, is necessarily dialogical in the sense of transhistoric – and more particularly *transgenerational.*

More on Dialogism, Restorative Justice and the Politics of Self-Portraiture as a 'Banlieue Writer'

Mounsi repeatedly describes himself, self-referentially, as a *writer*: 'My point *in this book* is neither to comfort nor to worry anyone but simply to give voice to a fraction of all things seen, experienced or observed' (9, emphasis added), he announces in the 'Foreword'. He also writes, 'It took me thirty years to move from the "Crimes and delinquency" section of the newspaper *to the literary section*' (21, emphasis added); '*I write because* human beings are, in the last analysis, a tale of agony – agony in a literal sense, the Greek sense of "struggle" – struggle with words and through words' (71). This abundance of self-characterization harks back to Mounsi's initial paradox. We have seen that Mounsi has a paradoxical relationship to the French language, in that he, on the one hand, defended the assimilation of immigrants into the French polity through their assimilation of said French language, while, on the other hand, he also values variations in their own right, i.e., 'accents'. He was thus both participating in the archipolitical phenomenon *and* resisting it.

Such claims, it can be argued (pro- and anti-archipolitical claims), had been made from the point of view of an abstract, general, disembodied relationship with the 'French' language. The same paradox continues to be very salient when it comes, for Mounsi, to discuss his status as a *banlieue* writer. Defining himself not only as a banlieue writer, but also as either an *archipolitical banlieue writer* or, on the contrary, as an *anti-archipolitical* banlieue writer appears to be very political business (in Rancière's sense of 'political'). Quite strikingly, there are passages in which Mounsi readily describes himself as fully archipolitical, and others when he, on the contrary, affirms his anti-archipolitical difference. It thus seems that an ambiguous definition of postcolonial banlieue literature emerges, where such literature, *in order to exist,* would need to be archipolitical to some extent, while also, to another extent, needing to be absolutely, resolutely anti-archipolitical; otherwise it would no longer be *banlieue* literature.

One way in which Mounsi defines postcolonial banlieue literature as archipolitical is by presenting it as completely dialogical with canonical 'French' literature. Mounsi thus narrates (60 *et passim*) the birth of his own vocation as a writer by relating it to his discovery, while imprisoned for delinquency, of the verses of the famous French medieval poet, François Villon. Villon, for his part, is said to have composed his *Ballade des Pendus* while also in prison, awaiting judgement and possibly execution after committing crimes, which is a detail likely to have drawn Mounsi to such reading at that particular time. This parallel between Villon and himself is made especially explicit in *Territoire d'outre-ville*: '[As I re-read Villon's *Balade des Pendus*], each word rolls out of my mouth *in exactly the same way*, I am sure, as it did come out of Villon's heart' (124, emphasis added). The commonality of Mounsi and Villon's shared experience with (transhistorical) Paris and more specifically with the immutable yet ever-changing Seine River leads Mounsi to further reflect on the creative writing process: 'I never stop wanting to understand this territory where I lose myself and find myself again. In the calm space of the white page, there is no trace of violence in my entire being any more', adding: 'The grey water of the Seine River flows in my eyes, washing away the entire city of Paris' (124).

With the sentence, 'In the calm space of the white page, there is no trace of violence in my entire being anymore', it does seem in effect that the previously discussed 'shout' has disappeared, as though the magnetic Parisian archipolitics had won and the centrifugal anti-archipolitics had been tamed, thanks to some would-be therapeutic virtue of literature.

More broadly still, Mounsi interestingly transfers the aforementioned postcolonial violence from 'reality' to literature. He *inscribes* it into literature, and more specifically into what he portrays as a distinctly 'French' literary tradition – thus, in a way, *circumscribing* and neutralizing this postcolonial violence. This movement, begun with the simile between himself-the-delinquent and Villon-the-criminal, finds its fulfilment through the explicit comparison between immigrant 'violent acts' and *very 'French'* nineteenth-century *literary* (novelistic) behaviour – with implicit references to masterpieces of 'French' literature, such as Jean Valjean's stealing of bread and struggle with Javert, his police nemesis, in Victor Hugo's *Les Misérables*: 'The links between young Arabs and the judicial institution are, in many ways, reminiscent of nineteenth-century novels' (59). In this regard, Mounsi first begins with a general statement in the third person: 'For a child who grew up in a ghetto, the first rebellion always expresses itself through crime. Extreme poverty [la misère], this old proletarian disease, is what governs all his actions' – after which he dramatically switches to the first person:

> I remember being hungry [....] I stole and had my reasons for it. I do not see however why I should spend time explaining why I stole – the reality of this world provides ample explanation for it. Honesty too is a luxury.
> (Mounsi 59)

The pathos-inducing themes of child poverty and child hunger are consequently vividly depicted both as a staple of Western (think Dickens) and French literature *and* very real contemporary socio-economic realities. Both are presented on an equal, antiracist, compassionate footing.

That said, even logically sound, well-meaning sentences meant to refute racist interpretations of crime numbers (in French media discourse, 'les chiffres de la délinquance') trying to show a link between 'foreignness of origins' and criminality, involuntarily end up reducing the singularity of postcolonial immigrant experience, which Mounsi has taken great pains to demonstrate through his two-generation theory: 'Of all life stages, adolescence is the most complex, incomprehensible one. Adults, although they were teenagers once, will never be able to really understand a teenager. Folly has always been, throughout human history, a characteristic of the youth' (58). This move, however, makes strategic sense; given the universality of the concept of youth, incidentally confirmed by contemporary scientific research which shows that the human brain is not fully formed until 'the mid-to-late 20s' (NIH),[10] there is no reason why, according to Mounsi, 'les jeunes', in media discourse should be a politically correct way of referring to *les jeunes Arabes*. The youth is the youth – totally and absolutely the *French* youth, regardless of origins. In doing so, Mounsi connects the perpetrators and the victims through their core shared a-political humanity, thus reaching an important first step in the theory and practice of restorative justice (Cook et al. 2006).

Furthermore, integration into the 'French' archipolity through literature (both the reading of literature and, in the case of *banlieue* writers, the writing of literature) is presented as an important solution to postcolonial problems. Mounsi indeed sounds here most conventional – understandably, since 'equality' is missing in the first place – in his defence of the 'French' *equalizing* public educational system:

> For a long time, I lived in a world devoid of words – a world of illiterates. The banlieue did give me slang, this middle term between thug language and French language, actually half-way between the insult and actual language.
> (70)

Much along the lines of Azouz Begag's *Les voleurs d'écritures* and *Les tireurs d'étoiles*, two short stories published just a few years before

Territoires d'outre-ville and centred around the 'magical', transformative value of public libraries, Mounsi states, 'Reading was the only durable *act of resistance* which I could oppose to this society during my youth' (70, emphasis added).

Therefore, reading and assimilating canonical French literature, which could have been interpreted in another context as an act of compliance vis-à-vis the archipolitics, is here felt by Mounsi as an 'act of resistance' and actual transgression:

> We must fight, with words, in order to demystify the strength of prejudice [....] *The true transgression*, for the ghetto youth, *is to appropriate themselves language*. Granted, it is with their claws that they are fighting right now, and with both hands. But the time of fingers will come – fingers to write, paint, sculpt, and draw! All human beings do have a form of culture; however violence is deemed to be the only language which [banlieue youth] knows. Is it right to assume that these children are all nature and have no culture, and will never have?
>
> (81, emphasis added)

This quotation takes up again the idea that young immigrants may *today* express themselves through 'violent acts', but will *tomorrow* use 'words' and 'meaning'. However, with the use of the word 'claws', Mounsi is, again, somewhat paradoxically and dangerously flirting here with right-wing ideas of 'immigrant youth' being 'des petits sauvages'[11] (which, in English, can be translated both as 'little rascals' and 'uncivilized savages', depending on the utterer's intention) and relies on notions of horizon and perfectibility which might not be satisfactory to voters voting in the here-and-now of political (non-humanistic) urgency. Significantly, though, Mounsi broadens his scope, as we not only pass from violence to written expression, but to *artistic* expression at large. In this regard, the 2011 film *Fracture*, directed by Alain Tasma, remains extraordinarily relevant to watch, as it powerfully exemplifies and brings to the screen to life – Mounsi's phrase 'fingers to draw instead of claws to fight' in the context of the *banlieue* – with shared humanity ultimately prevailing over extremely grave centrifugal (e.g., French-Jewish/French-Muslim) societal divisions.

Conclusion: Automation with (No) Revolution?

Postcolonial banlieue writers and artists such as Abd al Malik and Mounsi have displayed strategic ties with leftist French political philosophers like Stéphane Hessel (founder, in 2010, of *le mouvement des Indignés* akin to the Occupy Wall Street movement) and Alain Badiou (author, in 2011, of *Le Réveil de l'Histoire*, written in the aftermath of the Arab Spring, general European political instability and the 2008 financial crisis), very much interested in the idea of anti-capitalist revolution. The latter book is a criticism against all those 'satrapes-nigauds' (10),[12] as Badiou calls – satirically but very seriously – the politico-economic elite ruling over the modern world. Because, following Karl Marx's analysis, the choice in modern societies is between communism and barbaric capitalism, 'communisme ou barbarie' (25), capitalistic political systems may be found, according to Badiou, in only three possible states of necessary unrest: (1) 'émeute immédiate', riots immediately following an incident, yet not guided by what Badiou calls 'une Idée'; (2) 'émeute latente', riots in search of the communist idea; or (3) 'émeute historique', successful riots leading to actual historical changes, hopefully leading towards an ever-improved realization of the communist idea, in a sort of Hegelian dialectic spiral movement (29, 44, 53). For Badiou, 'incessant modernization' (10), or, as one might say, automation, which is fundamentally brought about by the constant international economic competition due to capitalism (Badiou, 10), is inseparable from the idea of revolution.

The alliance between immigrant workers and leftist intellectuals, while mentioned in *Territoire d'outre-ville*, is however far from obvious and even questioned, as Mounsi recounts several scenes he witnessed in which his father's coworkers (described as 'sous-proletariat', 49) expressed mistrust towards leftist revolutionaries. In one – in retrospect, quite humorous – passage which Mounsi situates shortly after the May 1968 strikes, the disconnect between the two social groups is so stark that the immigrant workers actually mistake the activists for police officers: 'C'est des keufs, ces keumés' ('These dudes are cops', 48). While toppling capitalism was already unlikely then, Mounsi points out that it became even more so after the collapse of the USSR in 1991: 'All the doctrinal landmarks that had been framing and determining economic philosophy for more than a century in terms of liberalism vs socialism wavered and waned' (50). As a result, Mounsi notes the 'void' ('le vide') that presently characterizes industry workers: 'C'est le vide autour d'eux et le vide en eux' – 'void around them and inside them' (50). This theme of extreme emptiness results from the death of the idea of revolution, or, as Mounsi would say, 'collective dream' (50). What killed the idea of revolution, according to Mounsi, is automation. Note the fatalistic tone in: 'Since the era of assembly lines, *the proliferation of automation* associated with the computer revolution has replaced the industrial revolution' (50, emphasis added), leading to mass unemployment. While Mounsi acknowledges the reality of 'transformations', whether they are economic, political, moral and technological (51), revolution against capitalism seems out of the realm of possibility.

Badiou *et al,* of course, would strongly push back against this idea; I hope to have shown that for Mounsi, however, the cultural revolution taking place through the proven agency of second-generation immigrant postcolonial writers is a heroic departure from mono-discursivity regarding the 'nation-France', countering the automatic reproduction of at least *cultural* capital.

Notes

1 The quotation marks around the signifier 'French' or 'France' throughout this essay are here to indicate my understanding of Frenchness as an ever-evolving, never definitively defined sign/signified, akin to Roland Barthes' use of the concept of mythology. In fact, these quotation marks are meant to act as perpetual reminders that in a poststructuralist/postmodern view, nations (such as 'France') are fluid constructs and should be understood as the opposite of stereotyped (i.e., automatic/automated) processes in the etymological sense of 'stereotype' (infinite reproduction, through printing, of one solid original plate).

2 This is in line with the question famously raised by Spivak ('Can the Subaltern Speak'?) – a question partially answered by Frantz Fanon in his *Les Damnés de la Terre* (translated into English as 'The Wretched of the Earth'), which underscores the dehumanizing effects of colonization (qua capitalism qua automation).

3 This dehumanization and devouring already began, according to Mounsi, through the enlisting of colonial populations into wars that were not theirs (First and Second World Wars). See Mounsi's reactivation of the concept of 'soldat inconnu' (unknown soldier, whose *identity does not matter*) in the following quotation: 'How many times were Algerians, Moroccans, Tunisians, and Senegalese sent to the front! So, it appears that France, formerly, had started to accept all people, no matter where they came from, no matter their skin color; it had become infinitely tolerant in the choice of its martyrs. After all, it was better not to be picky; it had even come up with the nice concept of "unknown soldier" ["soldat inconnu"]. Liberté, Egalité, Fraternité' (54). These anonymous deaths, anonymous masses of dead colonial flesh dumped onto the altar of French nationalism, are transcribed by Mounsi by the phrase '*chair à canon*' – 'cannon fodder' (54).

4 'Forgetfulness seems to be, in this country, something like a national duty' (31) While Mounsi uses here quite plainly the simple word 'oubli' ('forgetfulness'), the more commonplace expressions in French philology are 'désir d'oubli' or 'volonté d'oubli', which hark back to both Althusser's 'power of interpellation' (collective > individual) and Schopenhauer's idea of volition (individual > collective).

5 See Hélène Jaccomard, 'Racaille vs. Flics? Who's to Blame for Criminality and Delinquency in Franco-Maghrébine (Beur) Fiction'?

6 According to the article by *Le Parisien*, the banlieue 'way of talking' ('accent des banlieues') – characterized by speed, a way of ending sentences perceived as sharp ('brutal' in French), an Arabic-influenced way of pronouncing the letter 'r', a fairly limited range of vocabulary and mannerisms that include the repetitive use of words such as 'ziva'! (verlan for 'vas-y'!) or 'wesh' – is routinely disqualifying in job interviews. For a 2014 news segment titled 'L'accent des cités, un frein pour trouver un emploi' (in English: 'Hood accent: An obstacle for finding a job'), journalists working for France 2, the French public broadcaster, met with a Director of Human Resources based in Saint-Ouen, north of Paris, who confirmed that 'wesh-wesh language tics' were common motives for application rejections because clients would feel threatened and afraid. This notion of 'language tics' interestingly pertains to automated language, and attempts

to break it. As the 'Perdre son accent de banlieue' online Reddit forum shows, participants in the forum tend to wander from solution to solution, hesitating between resorting to the services of a speech therapist ('orthophoniste'), taking a theatre class, or recommending each other a 2004 book by baritone singer Louis-Jacques Rondeleux, *Trouver sa voix,* which advises controlling one's breathing, enriching one's vocal timber, and enlarging one's vocal register. Regarding speech therapists, a Québécois website (Centre Mosaïque de Québec – Clinique d'orthophonie) recognizes that speech therapists may help, but also gives a series of advice for self-help 'because we realize that it may not be possible for everyone who has an accent to take advantage of the structure provided by professional speech therapist services': identify what is to be worked on; record oneself; speak slowly; look at oneself in the mirror; listen to the radio and watch TV; and: 'practice, practice, practice'. In France, speech therapists preferably work with patients with speech pathologies, which an accent is not, as Isabelle Cottenet, a self-employed Coach in French Pronunciation, states on her website, hence advocating for accent reduction to be left to coaches in French pronunciation. Google searches for 'coach en prononciation française Paris' yield, however, a majority of results targeting wealthier anglophone expats moving to France, not banlieue residents trying to improve their chances of finding employment. Furthermore, accent reduction is linked to glottophobia, or accent-based discrimination (Blanchet 2016).

7 The impersonality of discourse.

8 Again the impersonality of discourse.

9 See also Joy DeGruy's concept of PTSS (Post Traumatic Slave Syndrome) or Crystal Fleming's similar idea of 'afterlife of slavery' as explained in *Rise Up! How You Can Join the Fight Against White Supremacy.*

10 'Although the brain stops growing in size by early adolescence, the teen years are all about fine-tuning how the brain works. The brain finishes developing and maturing in the mid-to-late 20s. The part of the brain behind the forehead, called the prefrontal cortex, is one of the last parts to mature. This area is responsible for skills like planning, prioritizing, and making good decisions' (NIH)

11 As shown by Capucine Licoys in her article on the phrases 'ensauvagement' or 'sauvageon' in French political rhetoric, the terms have mainly been used in right-wing/far-right circles as early as the 1990s to designate either banlieue youth, immigrants more generally, or more recently also Muslim extremists (2023). However, the fact that, in 2023, President Macron's Minister of the Interior, Gérald Darmanin, employed the very phrase 'ensauvagement' even though he is supposed to be part of a 'centrist' government, seems to indicate a certain banalization of the term.

12 Untranslatable '*mot-valise*' pun made up of the word 'satrape' (administrator of Ancient Persian empires) and 'attrape-nigauds' (scam, trap for naïve people) to thus designate technocratic modes of governance. With this figure of speech, Badiou thus argues that capitalism is presented, to the people who are naïve and docile enough to believe it, as the only possible mode of governance.

References

Althusser, Louis. *Lenin and Philosophy and Other Essays*. Monthly Review Press, 1971.

Apollinaire, Guillaume. *Œuvres poétiques*. Edited by Marcel Adéma and Michel Décaudin, Gallimard, 1959.

Astier, Henri. "France Riots: Why do the Banlieues Erupt Time and Time Again?" *BBC News*, https://www.bbc.co.uk/news/world-europe-66071455.

Badiou, Alain. *Le Réveil de l'Histoire*. Éditions Lignes, 2011.

Barthes, Roland. *Mythologies*. Seuil, 1957.

Begag, Azouz. *Les tireurs d'étoiles*. Seuil, 1992.

Begag, Azouz. *Les voleurs d'écritures*. Seuil, 1990.

Blanchet, Philippe. *Discriminations: Combattre la Glottophobie*. Textuel, 2016.

Centre Mosaïque de Québec. Clinique d'orthophonie. "La réduction de l'accent: conseils de nos orthophonistes." https://www.centremosaique.ca/reduction-de-laccent-conseils-orthophoniste/.

Ciné-Club de Caen. "Ballet Mécanique." https://www.centremosaique.ca/reduction-de-laccent-conseils-orthophoniste/.

Cook, Kimberly, Kathleen Daly, and Julie Stubbs. "Gender, Race and Restorative Justice – Introduction." *Theoretical Criminology*, vol. 10, no. 1, 2006, pp. 5–7.

Commazi, Karine, A. Flieller, et al. "L'accent des cités, un frein pour trouver un emploi." *France 2,* https://www.youtube.com/watch?v=g3X3ojiCHaw.

Corbet, Sylvie, and Barbara Surk. "France to Lift State of Emergency in Efforts to Allow Political Dialogue in Riot-hit New Caledonia." *AP News*, 2024, https://apnews.com/article/new-caledonia-france-kanak-097b74f915c025c4a3fae8ca73f4e28a.

Cottonet, Isabelle. *Secure Your Pronunciation.* Orthophoniste ou coach en prononciation française? https://www.secureyourpronunciation.com/blog/orthophoniste-ou-coach-en-prononciation-francaise/.

DeGruy, Joy. *Post-Traumatic Slave Syndrome: America's Legacy of Enduring Injury and Healing*. Uptone Press, 2005.

Fabula, la recherche en littérature. "Nouvelle parution: J.-M. Clerc, F. Cruveiller." *Parole de banlieue: Mounsi*, https://www.fabula.org/actualites/30114/j-m-clerc-f-cruveiller-parole-de-banlieue-mounsi.html.

Fleming, Crystal M. *Rise Up! How You Can Join the Fight Against White Supremacy*. Henry Holt and Company, 2021.

"Fracture." Dir. Alain Tasma. Perf. Samy Seghir and Anaïs Demoustier. France Télévision, 2010. DVD.

Gaquin, Audrey. "Une France plurilingue?" *The French Review*, vol. 79, no. 2, 2015, pp. 278–94.

Guène, Faïza. *Du rêve pour les oufs*. Hachette Littérature, 2006.

Henriques, Martha. "Can the Legacy of Trauma Be Passed Down the Generations?" *BBC*, https://www.bbc.co.uk/future/article/20190326-what-is-epigenetics.

Jaccomard, Hélène. "Racaille vs. Flics? Who's to Blame for Criminality and Delinquency in Franco-Maghrébine (Beur) Fiction?" *Hexagonal Variation: Diversity, Plurality and Reinvention in Contemporary France*, Ed. Jo McCormack, Murray Pratt, and Alistair Charles Rolls, Brill, 2011, pp. 157–76.

Le Boucher, Dominique. "Dialogue avec Mounsi." *L'Actualité Littéraire,* https://www.limag.com/Volumes/EmigrationCritiqueTt.pdf

Le Parisien. "Mais pourquoi les jeunes des cités ont-ils un accent?" https://www.leparisien.fr/archives/mais-pourquoi-les-jeunes-des-cites-ont-ils-un-accent-27-11-2012-2358671.php.

Licoys, Capucine. "'Ensauvagement' de la société: D'où vient cette expression utilisée par Gérald Darmanin?" *La Croix*, https://www.la-croix.com/france/Ensauvagement-societe-dou-vient-lexpression-utilisee-Gerald-Darmanin-2023-11-23-1201291965.

Mounsi. *Seconde Génération*. Les Disques Motors. CD, 1985.

Mounsi. *Territoire d'Outre-Ville*. Stock, 1995.

Mucchielli, Laurent. "'Insécurité,' 'sentiment d'insécurité': les deux veines d'un filon politique." *Après-Demain*, vol. 4, no. 16, 2010, pp. 3–6.
National Institute of Health (NIH). "The Teen Brain: 7 Things to Know." https://www.nimh.nih.gov/health/publications/the-teen-brain-7-things-to-know.
Rancière, Jacques. *La Mésentente. Politique et Philosophie*. Galilée, 1995.
Reddit. Online Forum. "Perdre son accent de banlieue." https://www.reddit.com/r/france/comments/yrn8hd/perdre_son_accent_de_banlieue/?rdt=34876.
Rondeleux, Louis-Jacques. *Trouver sa voix*. Seuil, 2004.
Schopenhauer, Arthur. *The World as Will and Representation*. Dover Publications, 1958.
Spieser-Landes, David. "Postcolonial Islam Thought and Rapped: Abd al Malik's *Révolution Pacifique* from within the French Nation." *Performing Islam*, vol. 4, no. 1, 2015, pp. 35–59.

8 Humans in the Loop as Post-Literary Ghosts

Discomfort and Disruption on Amazon Mechanical Turk

Bruno Ministro

Automation, Digital Labour and Microworking

Some technologies seem to work like magic. Artificial intelligence (AI) is one of them. Yet behind every technology lies a vast amount of hidden human labour. The people who provide this labour are not only Silicon Valley engineers, programmers, designers and sales teams; most come from the Global South and are referred to as 'ghost workers' by anthropologist Mary L. Gray and computer scientist Siddharth Suri. Data work demographics suggest a predominance of workers from East Africa, India and the Philippines. In recent years, Nairobi, Kenya, has become a global hotspot for recruiters.[1] The 'digital factory' (Altenried), with its dematerialized assembly line and large numbers of disposable workers, remains committed to making technology look like magic.

Asymmetric relations are not exclusive to today's technology. Research shows that the history of computing has always been embedded in racialized and classed social relations (Dyer-Witheford; Franklin; Dhaliwal; Whittaker). Throughout history, the tech industry has relied on the hard work of countless lower-skilled and lower-paid workers to build their empires, from punch-card operators in the 1950s to modern-day OpenAI contractors who annotate large volumes of data to train AI models. Online gig work through platforms such as Amazon Mechanical Turk has become even more prevalent with the emergence of new forms of digital labour over the last decade (Benanav; Jones).

Amazon Mechanical Turk (AMT) is an online crowdsourcing platform that connects businesses with remote workers who perform small, human intelligence tasks. These microtasks often involve data entry, image tagging, surveys or content moderation. Launched in 2005, AMT played a crucial role in the rise of microworking platforms and the broader gig economy. It was not just an early actor in this scenario; it was a significant driver in developing and popularizing microworking platforms. AMT was one of the first platforms to popularize the idea of microtasks: small, discrete tasks that needed human cognition but could be completed quickly

DOI: 10.4324/9781003545910-9

and often did not require specialized skills. The concept of the microtask then became the foundation for other forms of microworking – both technically and socially. By enabling businesses to break down larger projects into smaller, more manageable tasks that could be outsourced to what AMT described, and still describes, as 'a global, on-demand, 24x7 workforce', Amazon demonstrated how various companies could use such a model in different economic sectors. As tech critic Jaron Lanier explains, the whole process has 'a sense of magic, as if you can just pluck results out of the cloud at an incredibly low cost' (178).

Although microworking is just a small part of digital labour and the broader digital economy (Scholz; Fuchs; Jarrett), it is a particularly interesting phenomenon to look at in global capitalism due to its perpetuation of a global divide through invisibility and social inequalities.[2] On crowdsourcing platforms like AMT, workers perform their tasks as 'humans in the loop', a term in computer science and machine learning that underscores the interdependence of humans and technology in such systems. The concept of 'human in the loop', or even 'human computation', implies that human cognition is pivotal for developing or validating the output produced by machine learning and artificial intelligence models. By ensuring accuracy, addressing edge cases and enhancing system learning through the combination of human judgement with automated algorithms, human contributions lead to more reliable outcomes. Nevertheless, some scholars argue that such systems 'extract humanness' from workers in favour of the human-like automatism of the 'capture-machine' (Morreale et al.) or the 'extraction machine' sustained by the collective knowledge and efforts of humans (Muldoon et al.). AMT and similar crowdsourcing platforms have been used over the last two decades as frameworks to provide 'humans in the loop' for a vast array of tasks, including text transcription and image classification. For example, workers might annotate text and images to train large language models or teach self-driving cars how to detect obstacles, traffic lights, other vehicles, pedestrians and the road itself. In a sense, they are responsible for teaching *self*-driving cars how to drive. On the opposite side of the AI pipeline, workers might also be recruited not only to train new machine learning algorithms but also to evaluate the performance of unsupervised models.[3] This process effectively keeps them in the loop all the way down.

The innovation pressure felt today across different tech and social sectors has led to situations that, while extreme, are also somewhat amusing. For example, AI chatbots have proliferated in customer services in recent years, yet some services rely on AI impersonation. Without knowing, users often end up talking to humans who try to sound like chatbots to maintain the image of cutting-edge services. One such case involves a real estate chatbot called Brenda, as covered by Paris Marx for *Business Insider* in

2023. Brenda served as the initial point of contact for customers enquiring about apartment listings, providing basic details such as rental price and square footage. However, the chatbot responses often felt unnatural, and the system struggled with more complex inquiries, requiring human intervention to assist where the chatbot fell short. Again, automation shifts workers out of sight rather than erasing them. While new jobs are created, not erased, the quality of work can decline, leading to lower wages and more precarious employment. This is a loop as well, albeit of a different kind. We can consider a different case in Amazon's business ecosystem for a second example of innovation pressure and digital labour. The 'Just Walk Out' technology, introduced in 2020 at Amazon Fresh physical grocery stores, offers customers an AI-powered checkout-free experience. Though it seemed totally automated, it was later alleged that the system involved one thousand people in India tracking customers through a network of CCTV cameras to ensure accurate checkouts. Though these stories may seem humorous, when grouped together, the broader AI picture is more like a bad joke on automation. The digital labour associated with AI is, in fact, less about automation and more about poorly compensated, invisible work. This is also why some scholars, such as Hamid R. Ekbia and Bonnie A. Nardi or Antonio A. Casilli, have challenged the idea of automation and proposed the concept of heteromation instead. In this context, heteromation connects to AI impersonation, where human workers attempt to impersonate automated systems. Ultimately, heteromation is not a bug of AI; it is a feature.

Beyond the AI hype and innovation pressure, tech companies still need humans to complete small, repetitive tasks that computers cannot handle. The smooth functioning of services provided by big tech companies such as Amazon, Alphabet (Google), Meta (Facebook), Apple and Microsoft is only made possible by the hard labour of a large, unseen and intensely exploited workforce. One could argue that these individuals are not even employees, as they usually do not hold formal work contracts and are often paid per task, as is the case with AMT (see Williams et al.). An anonymous crowd of workers, colloquially known as Turkers or Turks, performs HITs – an acronym that stands for Human Intelligence Tasks – on Amazon's platform for mere pennies. Essentially, HITs are microtasks that a machine cannot complete automatically – a process that Amazon founder Jeff Bezos once referred to as '*artificial* artificial intelligence'. This description aligns with the name chosen by the entrepreneur for the platform, which reminds us of the machine invented by Wolfgang von Kempelen in 1770. The Mechanical Turk, or simply the Turk as it was called, was an automaton designed to appear as a chess-playing machine. It featured a life-sized figure in Ottoman robes seated behind a cabinet and captivated audiences across Europe and the Americas with its seemingly intelligent gameplay. Notable historical figures who played against the Turk include Napoleon Bonaparte and Benjamin Franklin. These high-profile matches helped cement the automaton's fame and mystique throughout the eighteenth and nineteenth centuries. With this, the invention both drew on and contributed to popular imaginaries of automation.[4] The Mechanical Turk was an automaton chess player that was, in fact, an elaborate hoax. The actual player was a chess master hiding inside a cabinet who could secretly move the pieces through levers and magnets. It worked like magic because, in reality, it was simply a magic trick.

Inspired by von Kempelen's Turk and using Amazon Mechanical Turk, in 2016, artist and programmer Lee Tusman created the *Literal Mechanical Turk*. As the title suggests, Tusman literally played with Turkers, paying $0.80 for each worker to make a single chess move. Workers were instructed to record themselves on video while performing the task. The final collection of videos shows the different contexts of the real people behind the computer screens. In Tusman's own words:

> In the response videos, it is possible to see into the abstracted-away black box of the Internet, to see through to the living rooms, offices, bedrooms and libraries of Turkers, to a diverse community of people worldwide. The process does not specifically create substantial equity for the participants [...] but it does work (I hope) to present an intimate view into labor on the Internet, and to humanize who is on the other end of the digital transaction.
>
> (Tusman)

Over the years, several artistic and literary works have critically engaged with Amazon Mechanical Turk to expose the unsettling truth about digital labour in global post-capitalism. Starting from Tusman's proposal to open an 'intimate view into labor on the Internet', I propose we look closely at a selection of some of these works, including projects by Aaron Koblin, Takashi Kawashima, Fred Benenson, Nick Thurston and Piotr Marecki. Additionally, I will comment on a research-creation project I developed myself, inspired and discomforted by some of these and other works. We might say I included myself as a human in the loop here, too.

I argue that digital labour platforms like Amazon Mechanical Turk, while often heralded as beacons of automation, in fact rely heavily on invisible, exploitative human labour. Through a critical examination of artistic and literary engagements with these platforms, I demonstrate how such works disrupt the narrative of seamless automation, exposing the underlying socio-economic inequalities and prompting a re-evaluation of the ethical implications of digital labour in our increasingly data-driven economy. Rather than close reading these literary and artistic works, I will look at them with a focus on the material conditions of their production. By material conditions, I mean, on the one hand, the work's media and creative processes of material inscription and, on the other, the social contexts and the formal aspect of labour relations implied in creating the works. In other words, I will look at the works and the workers in equal parts to understand what these works *are* and how they were *produced* to be as they are in their current form. Interestingly, their social implications are the same as those of other projects in tech business and digital development. In this way, these projects are not exempt from taking part in, and even from reinforcing, the exploitative systems they criticize. This also suggests that these pieces provoke a discomforting experience in those who interact with them, making the reading and fruition uncomfortable, as I will demonstrate throughout my analysis. With this, I argue that these projects are not only discomforting but also disruptive.

Discomfort on Amazon Mechanical Turk

In 2006, Aaron Koblin was responsible for what was one of the first creative artworks developed with the help of Turkers when AMT was little more than a novelty. The artist paid 10,000 workers $0.02 each to draw a sheep, resulting in a collection of drawings under the title *The Sheep Market*. Originally developed in Flash, various versions of the work were exhibited between 2006 and 2008 in Spain, Japan, the USA, Australia and the Netherlands. The presentations varied, sometimes displayed as digital pieces and other times as physical installations, including prints of the various drawings. The work's current webpage, ported into HTML5 after Adobe Flash became obsolete, features a clean, white background that gives the interface a minimalist appearance. At the centre of the screen is a large rectangular box that, when clicked, serves as a menu to access each of the 10,000 sheep. Each sheep animation appears frame by frame, as it was originally drawn by Turkers, in yet another box on the top. You can find amazing drawings (like sheep #4112 and #7049) and effortless ones where the Turker did not even try (sheep #9711 serves as an example). Yet, for Aaron Koblin:

> If every person drew a perfect sheep, they would all be the same and it would be a horrible project. It's actually seeing the ways that people fail, the different intricacies and character that comes from the individual that adds a lot.
>
> ('About The Sheep Market')

Each worker was simply instructed to 'draw a sheep facing left'. The webpage contains these and other details, such as the average time Turkers spent drawing each sheep, a calculation of their wage per hour and how many sheep were rejected. We are also informed, for instance, that the project was developed over 40 days, at a rate of about 11 sheep per hour. The fun nature of the project might be the only reasonable explanation for why Koblin detected 7,599 unique IP addresses for 10,000 drawings, meaning that as many as 2,400 people contributed twice or more. The work is designed to be interactive, allowing users to select and observe the sheep being drawn. Interactivity is central to the experience, inviting viewers to engage with the piece and watch the animations rather than just navigating a static image collection.

The project is reminiscent of a moment in *Le Petit Prince*, by Antoine de Saint-Exupéry, in which the narrator, an aircraft pilot, finds himself isolated in the Sahara Desert and hears a voice of a little man suddenly say:

> 'If you please – draw me a sheep...'
>
> When a mystery is too overpowering, one dare not disobey. Absurd as it might seem to me, a thousand miles from any human habitation and in danger of death, I took out of my pocket a sheet of paper and my fountain pen. But then I remembered how my studies had been concentrated on geography, history, arithmetic and grammar, and I told the little chap (a little crossly, too) that I did not know how to draw. He answered me:
>
> 'That doesn't matter. Draw me a sheep.'
>
> (16).

As it turns out, this little man was the Little Prince. After drawing some sheep – all rejected by the Prince – the narrator finally draws a relatively simple box and tells the boy that the sheep is inside. To his surprise, the Little Prince is delighted with this drawing. Delights and textual contaminations apart, Koblin's request can be read less playfully. Even the connections between the work's title, *The Sheep Market*, and the description of Amazon Mechanical Turk as a marketplace for human labour are somewhat dark and uncomfortable, to say the least.

In another project, Aaron Koblin, now in collaboration with Takashi Kawashima, asked Turkers to copy a tiny, abstract drawing provided by the artists, paying them $0.01 for the task. Workers did not know any details about the project or what the results would look like.[5] When all the drawings were combined, the result was a 100-dollar bill. The total labour cost to create the representation of the bill matched the value of the bill. Developed from November 2007 to March 2008, *Ten Thousand Cents* relied on custom drawing software created by the artists to record and assemble the drawings. The website initially featured a magnifying glass tool that allowed users to see each drawing in detail with a click. Today, this functionality is no longer available, but that is just a challenge of preservation. Back in 2007, the artists faced other challenges, as Kawashima explains in an interview ('AD-LITE' 116), they were uncertain about the final outcome until all the data was collected, making it challenging to remain confident throughout the five-month process. Their persistence was ultimately rewarded when they rendered the complete image after gathering all 10,000 drawings. Neither workers nor artists could anticipate the end results. Turkers did not know because they were just instructed to copy a rather abstract picture, a detail of the big picture. Artists could not know because it depended entirely on the materials the Turkers provided. This is an interesting aspect of this work because it illustrates that its creation was really processual, challenging and somewhat uncomfortable for the creators.

A different AMT project brings us closer to the literary domain. Published in book form, *Emoji Dick; or The Whale* (2010) consists of the translation of Herman Melville's *Moby Dick* into emoji. Every sentence in Melville's novel was translated three times by workers on Amazon Mechanical Turk. Another group of workers then voted on these translations, and the most popular version of each sentence made it to the book. Fred Benenson paid each worker $0.05 per translation and $0.02 per vote. The publication was possible due to both crowdsourcing and crowdfunding, with the $3,500 needed to produce it coming directly from a Kickstarter campaign.[6] Besides the details contained in an initial paratext ('About this book' vii), in the acknowledgements on the book's final pages (Benenson, *Emoji Dick* 723–735) there is a long list of names, including the Kickstarter supporters, family, friends and AMT workers' platform IDs, too. In this book, Benenson plays the role of editor and compiler, as is expressly made clear in all references included in the book's paratexts. The artist produced the book as sort of a data curator whose principal function is to conceptualize the idea, manage the AMT project, gather the information generated in the process and design the final product. The notion of authorship is challenged,[7] and so is the role of AMT workers as co-authors, given that they are mentioned as translators. A quick example of Benenson's collaboration with AMT, which cannot be fully conveyed in textual form, is the translation of Melville's original sentence, 'It's the unnatural combat of the four primal elements', by Turkers into a sequence of emojis: a skull, flexed biceps, fire, a cloud, an umbrella with rain, and a maple leaf (*Emoji Dick* 27). Writing about the project, in 'Emoji Dick and the Eponymous Whale', Lisa Gitelman comments that '*Emoji Dick* is a joke, a big one and timely' (200), detailing that 'if half the fun of *Emoji Dick* is indeed saying it, then the other half is not reading it' (205).

Even if it is a humorous and playful work, it was not created only for the sake of fun. As Benenson then explained in an interview, he was interested in exploring how digital technology shapes language, communication and culture, viewing emoji as either a decline or an evolution in this process. To address collective concerns about the future of human expression – particularly in social platforms – he chose to reinterpret a great work of literature through what it calls 'a strange new filter' ('The Revolution Will Be Crowdsourced (and Cute)'). One could say that this strange new filter deals mainly with the fact that language today does not exist as 'plain text', as was the case in some previous technologies. Multimodal writing includes a combination of text, images and other symbols that are algorithmically processed and distributed. In this context, translating a work of literature into everyday emoji is challenging and ultimately impossible, a process of rendering the book unreadable. Thus, we as readers get lost in translation, mirroring the experience of Turkers on AMT while working on this fun yet discomforting literary project.

In the blurb for *Of the Subcontract. Or Principles of Poetic Right* (2013), Nick Thurston describes his book as 'a collection of poems about computational capitalism' that 'reduces the poetic imagination to exploited labour and, equally, elevates *artificial* artificial intelligence to the status of the poetic' (Thurston 2013). This description finds no paradox in the procedures applied to generate the book, given that every poem was subcontracted to a Turker. The collection includes 100 poems numbered from 0.01 to 1.0 and organized according to the cost of production into four parts. Each poem features a header displaying a title, the time it took to write it and the average wage calculated per hour. These poems have a great diversity of writing styles, as exemplified by their titles, which range from idealistic and formulaic poetic titles such as 'Dreaming of Dreams', 'The Meaning of Life', 'Pillow Full of Tears' and 'Precious You' to more raw (and humorous) ones like 'Crazy About Cricket', 'A Poem You Did Not Write' and 'But Alas I Am Alone and Poor'.

Here is an integral reproduction of 'A Poem You Did Not Write' (poem number 0.17 on page 38), which took the Turker 03:26 minutes to complete and was paid $2.98/hr:

> To write a poem for you
> That would surely not do
> For you to take it and make it your own.
>
> I'd spend all my time
> For just a little more than a dime
> Giving you line after line of my rhymes.

Rhymes are not free
But they come easy to me
As I believe that by now I have shown.

These rhymes are for you
Do whatever you do
And fake it to make it your own.

Asked for a poem, a Turker chooses to write on the theme that is most immediate: the very act of writing on the platform and the entire crowdsourcing process. The tone is sarcastic and slightly bitter, with the worker suggesting that their time is being exploited solely for the benefit of the contractor. The issue, one might extrapolate, is not merely the relations of production but the relations of power. The Turker has no problem selling their time for project execution, understanding that all relations of production involve the appropriation of surplus labour from direct producers. However, there is discomfort in the fact that this appropriation allows the contractor to claim the poem as their own, faking authorship and expropriating the original creator for 'little more than a dime'.

Perhaps counter-intuitively, on this occasion, I want to pay special attention to the foreword rather than the poems in *Of the Subcontract.* In the foreword, McKenzie Wark praises Thurston's book and comments on a selection of poems in some detail. For Wark, '*Of the Subcontract* is a celebration of the changing seasons and the beauty of love' (9), also noting that '[n]owadays, however [...,] people prefer to read about computers, information technology and love' (16). On a more personal note, closer to the end of the foreword, Wark takes the opportunity to comment on some of their previous books, stating, for example, that '*A Hacker Manifesto* (...) is a remarkable and beautiful book that I wrote' (17) and its first section is 'very impressive' (18). As it turns out, the foreword was also outsourced, this time to a ghostwriter in Pakistan through Freelancer .com, with the ghostwriter committing to writing the text as if they were Wark. Additionally, the narrative makes it clear that the ghostwriter is fully aware that this is not a book of poems authored by Thurston. For example, there are several references such as: 'Thurston gave $0.40 to the poet for writing this poem on his behalf' (13). In this, the writer maintains a casual tone, as if it were the norm, making the experience of reading the foreword even more uncomfortable. After reading just the first page, no reader familiar with McKenzie Wark's writings would believe that they wrote it, even if this is only explicitly revealed in Darren Wershler's concluding essay. However, at some point, we cannot be sure whose voice this is:

> I wrote many books in my life and want to help people to share that same experience. I believe that there is a great writer inside everyone. Freelance work has given me an opportunity to write other's concepts, ideas and stories the way they want to write but they cannot [....] People hire other personnel to ghostwrite their books not only because it is in fashion but also they have something special to say but do not have time to express so.
>
> (Wark 14–15)

At this point, the reader may not know if the ghostwriter is still trying to impersonate Wark's voice or is now speaking for themselves – the latter seems more likely. In fact, it appears to be a little like a pitch for their own freelance services. Moreover, this quote offers a precise view into a part of the text where the authentic voice of the 'ghost' emerges and then vanishes again soon after. Ultimately, this might be the best way to describe the paradox of ghostwriting in the contemporary commodified world: 'they have something special to say but do not have time to express so' (Wark 15).

In *Wiersze za sto dolarów* (2017), which translates as *One-hundred Dollar Poems*, Piotr Marecki did something similar to Thurston by asking Turkers to 'write a modern poem in Polish'. He then assembled the results and published them in book form under his name. Marecki did not edit the poems in any way. The poems were arranged according to the amounts paid for creating each, ranging from $0.22 to $10, for a total of $100. In a technical report, the Polish writer described the different stages of the project in detail, including his first failed attempts to develop it as an experimental translation of Thurston's book with the assistance of Turkers ('Crowdsourcing and literature' 81–2). This experiment was later abandoned due to the lack of commitment, which researchers have recently described as a radical shift in AMT into low-quality and unusable data (Marshall et al.). In the report, Marecki refers to what came to be his definitive work as 'a literary experiment undertaken in the digital media environment (82). Following conceptual writer Kenneth Goldsmith's cue on 'uncreative writing', Marecki argues that '[w]riting using AMT is a kind of digital writing characterized by uncreative strategies and co-authorship' (80). Marecki's narrative is also relevant in another aspect related to the broad genre of 'uncreative writing': it provides details on how difficult it was to guarantee outputs of any quality, describing all the back and forth needed to accomplish the project between February 2017 and June 2017. This narrative clearly shows that, on the one hand, supposedly automated and effortless projects demand good management, and, on the other hand, as we have already seen for Benenson and Thurston, the management of information or curatorship of data is the kind of role that authors usually play in literary works based on remixing and repurposing appropriated text, more than writing new words on a blank page.

One of the most exciting aspects of this 'co-authorship' arises when, in their poems and although following all the project's rules, Turkers start reflecting on their own writing and working conditions. Eventually, they did it with great humour and even mocked the project. Here is an example that appropriates and repurposes the instructions provided by Marecki in the project form:

Apr 7 2017, 09:25 PM PDT, AVMXY0VQQ9C6A, 0.22 $
NAPISZ WIERSZ PO POLSKU NAPISZ WIERSZ WSPOLCZESNY PO POLSKY.
WIERSZ NIE MOZE BYC NIGDZIE WCZESNIEJ PUBLIKOWANY. PRAWA
AUTORSKIE PRZECHODZA NA REQUESTERA. REQUESTER MOZE
OPUBLIKOWAC WIERSZ. WIERSZ NIE MOZE BYC SKOPIOWANY Z INTERNETU.
WIERSZ MOZE BYC NA DOWOLNY TEMAT.
NATIVE POLISH SPEAKER EQUAL TO 1
MASTERS HAS BEEN GRANTED.
Apr 7 2017, 09:25 PM PDT, AVMXY0VQQ9C6A, 0.22 $
WRITE MODERN POEM IN POLISH WRITE MODERN POEM IN POLYSH.
POEM CANNOT BE PREVIOUSLY PUBLISHED. THE COPYRIGHT IS
TRANSFERRED TO THE REQUESTER. THE REQUESTER CAN PUBLISH THE
POEM. THE POEM CANNOT BE COPIED FROM THE INTERNET. THE POEM
CAN BE ON ANY SUBJECT.
NATIVE POLISH SPEAKER EQUAL TO 1
MASTERS HAS BEEN GRANTED.

('Crowdsourcing and literature' 83)

It is worth returning to Nick Thurston's work to comment that the emergence of Turker's discourse is not strictly exclusive to Marecki's work. Besides the poem analysed previously, in the poem 'Am I Blind, Or Maybe Dumb'? (24), a worker writes:

Am I blind, or maybe dumb?
To see TWO cents has made me numb.

Would you do work for this measly amount?
Would you take it seriously, would it even count.

This is insulting in so many ways,
But it seems a trend, the newest craze.

I do not mind writing when the prices are right,
But two cents is insulting and not worth the fight.

No payment and a rejection are sure to come,
But I could not let this pass without saying

'I AM NOT DUMB'.

In yet another one, a quatrain simply titled 'Turk' (51), we read:

Here I sit, and I work,
Being a nasty little Turk,
I hardly get paid and the hours suck,
So give me a fucking buck.

These and other examples in Marecki's and Thurston's books can be connected to what Jamie Woodcock and Callum Cant call acts of resistance in Turkers' writing. By analysing Turkers' writing, in their case mainly blogs and articles reflecting and denouncing their working conditions as opposed to literary or artistic contributions, Woodcock and Cant argue that these pieces of writing represent 'experiments from workers about how to take back control' (159). The artists and writers whose works I have analysed so far are not engaged in similar, direct action. However, their works contribute to exposing the reality behind digital labour, raising awareness about microworking and reflecting on social challenges in a networked, post-digital world.

To return to Marecki, *Wiersze za sto dolarów* is described as a 'crowdsourcing project' that resulted in a 'traditional book' ('Crowdsourcing and literature' 82). While the book may be conventional in its material form, the fact is that, like Benenson's and Thurston's books, it is a post-digital book in the sense that it gathers and reassembles the contents that were dynamically generated *on* and *from* the network – the network of workers, the network of the internet, the networks of the network.

As a final example, I would like to discuss my own work, developed as a research-creation project. 'Progress in work' is a digital artefact created initially in 2021 and published for the first time in 2022 in *Electronic Literature Collection* vol. 4.[8] To develop the piece, I created two separate projects on Amazon Mechanical Turk: one for texts and another for images. The text project was based on fragments from Lawrence Venuti's book *The Translator's Invisibility*. I collected every epigraph from the book, split them to be more manageable and set all the tasks on the platform. Then, Turkers were commissioned to translate a fragment into their native language.[9] In the second project on the platform, I instructed

Turkers to take photos of their rooms or workplaces. By doing that, they exposed their privacy to a certain degree – enough for the final work to create a sense of discomfort. Turkers were paid $0.05 per task, and the project's final cost was $29.61 in total, which included $8.20 in fees charged by MTurk. All these details are featured in the work itself as a sort of uncomfortable metadata. I did reject some of the contributions; some texts were obviously machine-translated, and many images were taken from the Google search engine, which was not the point. Though I refused to accept some of these, I did, however, accept other texts and images just because I found them interesting in one way or another, even if they were taken from the web. In sum, I was unfair.

As I have already pointed out, platform workers are invisible because, on AMT, they are literally just numbers. There are no usernames, only randomly generated IDs. As thoroughly shown by previous research, AMT relies on worker invisibility by design (see Irani and Silberman). I wanted to make sure that Turkers were credited. However, all I had access to were their IDs on the platform. I included those on two separate tabs in the work's navigation menu and titled the respective lists 'Translators / Writers / Humans by Worker ID' and 'Photographers / Workplaces by Worker ID'. Ultimately, Amazon's Terms of Service declare that the copyright is mine because I paid for it. I, therefore, felt that I also had to take a photo of my desk, which appears in another tab along with the following copyright notice:

> Thanks to Amazon's Terms of Service, which state that all ownership and intellectual property rights will 'vest with that Requester immediately upon your performance of those Tasks,' the ownership of any work produced through the platform lies in the hands of whoever made the original work request.

By asking Turkers to take a photo of their rooms or workplaces, I tried to understand whether this would render them more visible. From this, we can at least agree that they exist. Sometimes, instead of taking a photo of their office or desk, they went for a selfie, which I accepted. These photos show us that these microworkers are not just '*artificial* artificial intelligence'. Nevertheless, I am still not convinced that this kind of project 'humanizes' these people – to use a term from Tusman. I think that, all in all, it makes everything uncomfortable.

As for the interface, the minimalist design of *Progress in work* aims to imitate the cold, grey interface of microtask management platforms. In addition, the piece functions like a mock platform where the user is supposedly paid per click. The text generator includes a 'hit' button, and the image generator includes a 'click' button. At the bottom of the page, a banner will sum the 'earned USD cents' for each click on those two buttons. In the 'terms and conditions' tab, you can read the following fake instructions:

> Terms of Service Agreement
>
> Earn money now by clicking on randomized buttons. Check the total amount earned at the bottom of your screen. When you are done, send an email to Amazon Mechanical Turk customer service to claim your money. Note that by navigating the website and reading this work of electronic literature, you guarantee you are not a robot.

This is important for the user experience because it places the reader in a similar position to the worker. Rather than reflecting on literary and artistic quality or seeing the works as closed artefacts on their own, *Progress in work* and the other experiments I have analysed here ask what it means to create content in the digital media environments of the twenty-first century. Moreover, they reflect on how this distributed authorship is glued to the division of labour in our datafied, post-digital world.

Disruption on Amazon Mechanical Turk

If data is the new currency in the age of commodified information (Wark, *Capital is Dead*), we must ask ourselves who is accumulating that capital. Today, there are many common questions regarding the data used in artificial intelligence and large language models: what data is this? Who generates it? Who manages it? And who owns it? We need to add yet another question: who 'owns' the people who process this data? The answer is the invisible workers taking care of annotating, cleaning and verifying this huge and troublesome amount of data.

Amazon's case is paradigmatic. Besides AMT, Amazon owns a substantial number of assets, from the e-commerce giant platform Amazon .com to 319 million square feet of warehouses in the United States alone. Less known but highly relevant in tech, the company provides a variety of services through its Amazon Web Services (AWS) programme, including cloud computing and APIs for other tech companies. Amazon was a major player in the dotcom era of the 2000s. Today, it is part of the so-called 'big tech', also known as 'the big five'. Starting as an online bookstore before branching out to other, multifaceted business sectors, Amazon has always been a tech enterprise at its core. Some even suggest that its business has always been more about gathering data than selling products (Packer). Over the years, poor working conditions at various Amazon warehouses across the United States and United Kingdom have repeatedly been brought into the spotlight in the news (Kantor and Streitfeld). In the introduction to *Amazon: At the Intersection of Culture and Capital*, editors Paul Smith, Alexander Monea and Maillim Santiago point out that Amazon has not only changed the face of business but has also had a broader cultural impact on society. In their perspective, Amazon's business model relies not only on its dominant market position and platform structure but, more fundamentally, on the exploitation of an expanding global workforce.

Amazon is what is called a 'disruptive' company in tech business jargon because of its constant cutting-edge innovations. The adjective 'disruptive' has been progressively appropriated in tech to mean groundbreaking or revolutionary solutions, describing innovations that significantly alter industries, challenge traditional business models and create new markets or value networks. I contend that we need to reclaim the negative connotations of the word 'disruptive' to restore its original meaning as something that disturbs and interrupts the order of events and social processes. Historically, art and literature have served as realms where language is both disrupted and disruptive. As generally accepted, language encapsulates all kinds of power dynamics. In our context, another terminological example proves this point. In 2005, when launching AMT at the Massachusetts Institute of Technology, Bezos explained it in these terms: 'You've heard of software-as-a-service. Well, this is human-as-a-service'. According to Lilly Irani, a scholar whose research focuses on AMT and digital labour, innovation also means invisibility and alienation:

> To serve is to make labor and attention available for they who are served; to promise service is to be bound, by duty or by wage, to the will of the served. Among computer scientists, 'as-a-service' builds off of this common sense meaning and more specifically indexes a division of technical labor by which programmers can access computational processing functions housed and maintained on the internet and by someone else.
>
> (13–14)

Looking back at the history of technology, Matteo Pasquinelli, in *The Eye of the Master: A Social History of Artificial Intelligence*, follows the invention of machinery through the lens of the division of labour, spanning from Charles Babbage's Difference Engine to Taylorism and the current forms of digital labour in artificial intelligence. As Pasquinelli summarizes in his reading of Karl Marx,

> it was not the invention of the steam engine (*means of production*) that triggered the Industrial Revolution (as it is popular to theorize in the ecological discourse), but rather the developments of capital and labour (*relations of production*) demanding a more powerful source of energy.
> (107)

By dislocating the focus from the means of production to the relations of production, this view contends that it was not the invention of machines and technology throughout history that changed how workers work; it was the other way around. This might be one of the reasons why several studies suggest that AI is creating more jobs than taking existing ones at present. There is no doubt that tech companies need many contractors to train AI models. Yet, a recent report (Gilardi et al.) also shows that large language models are getting better at jobs previously outsourced to crowd-workers on AMT and other platforms, with some outperforming human workers for text-annotation tasks. This may soon represent a new stage in automation. Again, this automation will not appear magically; it will result from the workers behind the machine. Once they have completed the extensive and tedious tasks of training AI, they will be discarded.

Digital labour, as I have demonstrated with the help of different artistic and literary experiments, is not only about first-world digital nomads. It is about exploitation, alienation and 'digital lethargy' spread across global networks. Tung-Hui Hu defines 'digital lethargy' as the fatigue, disillusionment and apathy that arise from living under digital capitalism. In the specific case of Amazon Mechanical Turk, these characteristics are not aberrations but are instead built into the system. Exploitation is constitutive of the way the digital platform works. The 'ghost workers' on the AMT platform are the 'ghosts in the machine', in the famous phrase of Gilbert Ryle popularized by Arthur Koestler. In the end, like von Kempelen's Turk, the magic of machines is just real people inside a black box.

The literary and artistic pieces I have analysed serve as examples of radical approaches that disrupt the inner mechanisms of that black box. By connecting the dots between their disruptive powers, I see them as examples of tactical media. For Rita Raley, 'tactical media signifies the intervention and disruption of a dominant semiotic regime' and it 'operates in the field of the symbolic', which is 'the site of power in the postindustrial society' (6). This is their symbolic contribution and their potential strength. We all know that these literary and artistic works will not solve anything; it would be impossible for them to do so. In fact, by dealing with digital labour, they even reproduce and reinforce its exploitative mechanisms. However, these experiments exploit the exploitation in a way that makes it become so exposed that it also becomes unbearable. In other words, nothing in these creative works will be helpful in the same vein as workers-led groups or projects like the Turkopticon[10] and the Data Workers' Inquiry.[11] Still, through their semiotic and symbolic languages, these works expose current material conditions and engage in alternative views for society at large.

The digital domain, often perceived as a realm of sheer innovation and seamless operations, conceals an intricate web of human labour that is indispensable yet frequently overlooked. As we delve deeper into the artistic and literary engagements with AMT, it becomes evident that these platforms, while enabling microtasking and the illusion of automation, perpetuate a form of digital labour that is both ubiquitous and invisible. The artistic interventions I have explored do not merely represent or reimagine the labour hidden behind the digital veil; they actively participate in and critique the very ecosystems that exploit this labour. By complicating the boundaries between machine and human, these works lay bare the socio-economic dynamics that fuel the digital economy, challenging us to reconsider the ethical implications of such systems. Through a blend of creative expression and critical insight, these projects illuminate the often unseen human elements that sustain and complicate our understanding of automation, digital labour and artistic production in the contemporary technological landscape.

Notes

1 Among several cases that have made the headlines in recent years, I would like to highlight two articles: one by Billy Perrigo for *Time* on ChatGPT and Kenyan workers (https://time.com/6247678/openai-chatgpt-kenya-workers/) and another, more comprehensive investigation by Josh Dzieza for *The Verge* and *New Yorker* magazine (https://www.theverge.com/features/23764584/ai-artificial-intelligence-data-notation-labor-scale-surge-remotasks-openai-chatbots).

2 Another example is content moderation. This is an entirely different job with various specific issues that arise from being paid to screen flagged content uploaded to social media sites. Nonetheless, it connects to the same reality of microworking, as thoroughly shown by Sarah Roberts in *Behind the Screen: Content Moderation in the Shadows of Social Media*.

3 For further discussion and critique of different worker roles, see the article 'The Trainer, the Verifier, the Imitation: Three Ways in Which Human Platform Workers Support Artificial Intelligence' (Tubaro et al.).

4 For a complete history, see Gerald M. Levitt's book *The Turk, Chess Automaton*.

5 For a similar blind strategy but in the musical realm, see also *Bicycle Built For Two Thousand* (2009), created by Aaron Koblin, Daniel Massey and two thousand Turkers singing *Daisy Bell*, the first example of speech synthesis dating back from 1962. https://www.bicyclebuiltfortwothousand.com/.

6 *Emoji Dick*'s webpage offers further details about the project: www.emojidick.com.

7 Traditional paradigms of authorship and their implications for contemporary literary creation have long been challenged, particularly by experimental literary forms that incorporate information technologies, including concrete poetry, conceptual writing and digital literature. Benenson's book can be viewed as a critical part of this self-reflective literary corpus, examining the nature of authorship itself.

8 My work in *ELC* vol. 4 is available at: https://collection.eliterature.org/4/progress-in-work.

9 Examples of epigraphs taken from Venuti's work include: 'How many people today live in a language that is not their own? Or no longer, or not yet, even

know their own and know poorly the major language that they are forced to serve?' (Deleuze and Guattari); 'The fundamental error of the translator is that he stabilizes the state in which his own language happens to find itself instead of allowing his language to be powerfully jolted by the foreign language' (Pannwitz); 'I see translation as the attempt to produce a text so transparent that it does not seem to be translated. A good translation is like a pane of glass. You only notice that it's there when there are little imperfections – scratches, bubbles' (Shapiro); 'Translation is defined as a second-order representation: only the foreign text can be original, an authentic copy, true to the author's personality or intention, whereas the translation is derivative, fake, potentially a false copy' (Venuti).

10 The Turkopticon is a software that provides a way for workers to rate requesters, and not just the other way around, as it is the default on AMT. https://turkopticon.net/ (See also Irani and Silberman's article on the Turkopticon).

11 The Data Workers' Inquiry is a community-based project that brings together data workers and data researchers to understand and organize forms of resistance in globally networked economies. https://data-workers.org/.

References

Altenried, Moritz. *The Digital Factory: The Human Labor of Automation.* U Chicago P, 2022.

Benanav, Aaron. "Automation and the Future of Work - I." *New Left Review*, vol. 119, Oct. 2019, pp. 5–38.

Benenson, Fred. *Emoji Dick, or, The Whale*. Lulu, 2010.

Benenson, Fred. Interview. "The Revolution Will Be Crowdsourced (and Cute)." By Sally Law. *The New Yorker*, 22 Sep. 2009, https://www.newyorker.com/books/page-turner/the-revolution-will-be-crowdsourced-and-cute. Accessed 11 Sep. 2024.

Casilli, Antonio A. "Waiting for Robots: The Ever-Elusive Myth of Automation and the Global Exploitation of Digital Labor." *Sociologias*, vol. 23, no. 57, 2021, pp. 112–33, https://doi.org/10.1590/15174522-114092. Accessed 11 Sep. 2024.

Dhaliwal, Ranjodh Singh. "The Cyber-Homunculus: On Race and Labor in Plans for Computation." *Configurations*, vol. 30, no. 4, Sep. 2022, pp. 377–409, https://doi.org/10.1353/con.2022.0028. Accessed 11 Sep. 2024.

Dyer-Witheford, Nick. *Cyber-Proletariat: Global Labour in the Digital Vortex.* Pluto Press, 2015.

Ekbia, Hamid R., and Bonnie A. Nardi. *Heteromation, and Other Stories of Computing and Capitalism.* MIT Press, 2017.

Franklin, Seb. *The Digitally Disposed: Racial Capitalism and the Informatics of Value.* U Minnesota P, 2021.

Fuchs, Christian. *Digital Labour and Karl Marx*. Routledge, 2014.

Gilardi, Fabrizio, et al. "ChatGPT Outperforms Crowd Workers for Text-Annotation Tasks." *Proceedings of the National Academy of Sciences*, vol. 120, no. 30, Jul. 2023, https://doi.org/10.1073/pnas.2305016120. Accessed 11 Sep. 2024.

Gitelman, Lisa. "*Emoji Dick* and the Eponymous Whale." *Book Presence in a Digital Age*, edited by Kiene Brillenburg Wurth et al., Bloomsbury, 2018, pp. 195–210.

Hu, Tung-Hui. *Digital Lethargy: Dispatches from an Age of Disconnection.* MIT Press, 2022.

Irani, Lilly. "The Cultural Work of Microwork." *New Media & Society*, vol. 17, no. 5, May 2015, pp. 720–39, https://doi.org/10.1177/1461444813511926. Accessed 11 Sep. 2024

Irani, Lilly, and M. Six Silberman. "Turkopticon: Interrupting Worker Invisibility in Amazon Mechanical Turk." *Proceedings of the SIGCHI Conference on Human Factors in Computing Systems*, ACM, 2013, pp. 611–20, https://doi.org/10.1145/2470654.2470742. Accessed 11 Sep. 2024.

Jarrett, Kylie. *Digital Labor*. Polity, 2022.

Jones, Phil. *Work Without the Worker: Labour in the Age of Platform Capitalism*. Verso, 2021.

Kantor, Jodi, and David Streitfeld. "Inside Amazon: Wrestling Big Ideas in a Bruising Workplace." *The New York Times*, 15 Aug. 2015, https://www.nytimes.com/2015/08/16/technology/inside-amazon-wrestling-big-ideas-in-a-bruising-workplace.html. Accessed 11 Sep. 2024.

Kawashima, Takashi. "AD LITE - Takashi Kawashima, Unleashed." *Creativity*, May 2008, p. 116.

Koblin, Aaron. "About The Sheep Market." *MIT – Docubase*, 2008, https://docubase.mit.edu/project/the-sheep-market/. Accessed 11 Sep. 2024.

Koblin, Aaron. "The Sheep Market." 2006, http://tobib.spline.de/sheep/. Accessed 11 Sep. 2024.

Koblin, Aaron, and Takashi Kawashima. "Ten Thousand Cents." 2008, www.tenthousandcents.com. Accessed 11 Sep. 2024. Accessed 11 Sep. 2024.

Lanier, Jaron. *Who Owns the Future?* Simon & Schuster, 2014.

Levitt, Gerald M. *The Turk, Chess Automaton*. McFarland & Co, 2000.

Marecki, Piotr. "Crowdsourcing and Literature: A Report from the Project 'Wiersze Za Sto Dolarów.'" *World Literature Studies*, vol. 9, no. 3, 2017, pp. 79–86, https://wls.sav.sk/wp-content/uploads/WLS3_2017_Marecki.pdf. Accessed 11 Sep. 2024.

Marecki, Piotr. *Wiersze Za Sto Dolarów*. Korporacja Ha!art, 2017.

Marshall, Catherine; et al. "Who Broke Amazon Mechanical Turk?: An Analysis of Crowdsourcing Data Quality over Time." *WebSci '23: Proceedings of the 15th ACM Web Science Conference 2023*, p. 335–45. DOI: 10.1145/3578503.3583622

Marx, Paris. "Artificial Intelligence's Dirty Secret." *Business Insider*, 12 Feb. 2023, https://www.businessinsider.com/chatgpt-ai-will-not-take-jobs-create-future-work-opportunities-2023-2. Accessed 11 Sep. 2024.

Ministro, Bruno. "Progress in Work." 2021, https://hackingthetext.net/progress-in-work/index.html. Accessed 11 Sep. 2024.

Morreale, Fabio, et al. "The Unwitting Labourer: Extracting Humanness in AI Training." *AI & Society*, May 2023, https://doi.org/10.1007/s00146-023-01692-3. Accessed 11 Sep. 2024.

Muldoon, James, et al. *Feeding the Machine: The Hidden Human Labor Powering A.I.* Bloomsbury, 2024.

Packer, George. "Cheap Words." *The New Yorker*, 9 Feb. 2014, https://www.newyorker.com/magazine/2014/02/17/cheap-words. Accessed 11 Sep. 2024.

Pasquinelli, Matteo. *The Eye of the Master: A Social History of Artificial Intelligence*. Verso, 2023.

Raley, Rita. *Tactical Media*. U Minnesota P, 2009.

Roberts, Sarah. *Behind the Screen: Content Moderation in the Shadows of Social Media*. Yale UP, 2019, https://doi.org/10.2307/j.ctvhrcz0v. Accessed 11 Sep. 2024.

Saint-Exupéry, Antoine de. *The Little Prince*. Translated by Katherine Woods, Harcourt Brace & Company, 1995.

Scholz, Trebor, editor. *Digital Labor: The Internet as Playground and Factory*. Routledge, 2013.

Smith, Paul, et al., editors. *Amazon: At the Intersection of Culture and Capital*. Rowman & Littlefield, 2023.

Thurston, Nick. *Of the Subcontract. Or Principles of Poetic Right*. Information as Material, 2013.

Tubaro, Paola, et al. "The Trainer, the Verifier, the Imitator: Three Ways in Which Human Platform Workers Support Artificial Intelligence." *Big Data & Society*, vol. 7, no. 1, Jan. 2020, https://doi.org/10.1177/2053951720919776. Accessed 11 Sep. 2024.

Tusman, Lee. "Literal Mechanical Turk." 2016, https://leetusman.com/projects/literal-mechanical-turk/. Accessed 11 Sep. 2024.

Venuti, Lawrence. *The Translator's Invisibility: A History of Translation*. Routledge, 1995.

Wark, McKenzie. *Capital Is Dead: Is This Something Worse?* Verso, 2019.

Wark, McKenzie. *"Foreword: Earn Money Just by Writing Your Mind". Nick Thurston, of the Subcontract. Or Principles of Poetic Right*. Information as Material, 2013.

Whittaker, Meredith. "Origin Stories: Plantations, Computers, and Industrial Control." *Logic(s) Magazine*, 2023, https://logicmag.io/supa-dupa-skies/origin-stories-plantations-computers-and-industrial-control/. Accessed 11 Sep. 2024.

Williams, Adrienne, et al. "The Exploited Labor Behind Artificial Intelligence." 13 Oct. 2022, https://www.noemamag.com/the-exploited-labor-behind-artificial-intelligence/. Accessed 11 Sep. 2024.

Woodcock, Jamie, and Callum Cant. "Platforms, Resistance, Organizing." *Amazon: At the Intersection of Culture and Capital*, edited by Paul Smith et al., Rowman & Littlefield, 2023, pp. 149–63.

9 Bricolage, Wild Thought and the Automation of Knowledge

Madeleine Chalmers

Tinkering, DIY, MacGyvering, jerrybuilding, cobbling together, bodging, patching up: the French term 'bricolage' has many potential translations, spanning amateur hobbies and daringly innovative solutions, half-baked projects and dubious fixes. The term 'automation', on the other hand, conjures up a different set of associations: processes requiring minimal human intervention, engineered and optimized for maximum efficiency and productivity. However, in this chapter, I propose that bricolage offers resources for thinking about the automated knowledge at stake in Large Language Models (LLMs) such as Chat-GPT, Claude or Gemini.

Bricolage has a life far beyond the workshop, with applications in avant-garde art, entrepreneurship and even genetics (Johnson 2012). In almost every case, the term is traced back to the definition provided by the French anthropologist Claude Lévi-Strauss in his 1962 work, *Wild Thought*: '[t]he *bricoleur* remains someone who works with his hands, using means [...] skewed in comparison with those of the professional craftsman' (20). The bricoleur combines existing elements in unexpected and unorthodox ways: to repair, build or replace. Lévi-Strauss places the bricoleur in the same epistemological tradition as the myth-maker, who weaves narratives from events in ways that are 'skewed' compared to the analytical scientist (*WT* 25). Indeed, the bricoleur and myth-maker sit in opposition to the engineer and scientist, who create 'events [...] thanks to structures' of their own design, reversing the dynamic of the bricoleur-myth-maker (Lévi-Strauss, *WT* 26).

Yet while we might readily associate technological innovation with the practice of engineering, bricolage is threaded through the history and mythology of computing and information systems. Alexander Galloway has drawn out the relationship between the Industrial Revolution and the history of computing, unpicking a persistent myth which binds together the 'first computer' and Jacquard's automated loom. This *locus classicus* is rooted in Ada Lovelace's observation that the Analytical Engine she designed (but did not actually build) with Charles Babbage '*weaves*

DOI: 10.4324/9781003545910-10

algebraical patterns just as the Jacquard loom weaves flowers and leaves' (696). However, as Galloway observes, 'Jacquard did not invent a new loom so much as rediscover and recombine existing techniques into a new system' (*Uncomputable* 74). Ellen Harlizius-Klück (on whom Galloway draws extensively) sets the real origins of computing in 'the cheap, self-made [...] mechanisms with untranslatable names invented and used by nameless regional weavers' (192). In this instance, the bricolage of myth occludes the 'real-life' bricolage of computing.

At other moments, bricolage and mythology have worked in partnership. In the 1970s and 1980s, AI working cultures were characterized as 'neat' or 'scruffy' (Hiesinger 83). Innovation was framed with well-worn Hollywood tropes of cool-headed, methodical and logical engineers, pitted against the 'scruffy' maverick hacker whose unorthodox methods outwit systems to yield dazzling breakthroughs. In 1995, Sherry Turkle explicitly identified practices of hacking as 'virtuoso' performances in which coding 'bricolage' became high art (51–4). A decade later, the information systems theorist Claudio Ciborra described bricolage as epistemology, stylish flourish and rebellion (47). Indeed, bricolage has also been aligned with concepts of remediation (Grusin and Bolter 1999) and the individual 'remixing' of elements in internet culture (Deuze 2006). These theorists all place Lévi-Strauss's definition as the point of origin for their own usage. To describe the automated knowledge produced in interactions with LLMs as a process of bricolage would therefore appear to be merely the next episode of this mythos.

However, there is one obvious and crucial difference between this form of bricolage and that enacted by LLMs. In the context of this chapter, LLMs are to be understood in general terms as models trained on vast datasets drawn from the internet, journalism and other sources, used to power chatbots with which users can interact. The LLMs at work in generative AI are autoregressive, transforming previous outputs into new inputs within each temporary interaction. However, they require the action of a human prompter, and the conversations which we have with them do not add to their store of data. In reassembling existing text or images based on recognized patterns and calculations of probability, an LLM is both bricolage – an accretion of training data, testing, corrections and recursions – and bricoleur, working in dialogue with the human. In this chapter, I do not intend to address the complexities of determining what constitutes 'agency' or 'intelligence'. Renovating Lévi-Strauss's term allows us to move beyond debates over whether LLMs are mere 'stochastic parrots' (Bender et al.). Rather than asking who or what is thinking, and whether its operation can indeed be described as 'thinking', the concept of bricolage allows us to focus on questions of process and output: what kind of automated knowledge is being produced and how do we interact with it?

In addition to this processual focus, the connection which Lévi-Strauss draws between the material world of bricolage and the narrative world of myth also allows us to reflect on the entanglement of those terms in our own lay relationship to this technological milieu. This is a confection of narratives, metaphors and histories which draw on our oldest myths (see Cave et al.; Cave and Dihal). As we seek to keep pace with developments which outstrip our conceptual frameworks, we find ourselves reconfiguring those we do have to accommodate the seemingly new. Those narratives, metaphors, histories and hypes in turn shape the kinds of technological thinking and making that are possible. We refer to AI, algorithms, computation, but these signifiers have a tendency to float, applying equally to Nobel Prize-winning research into protein folding ('AlphaFold') or to a student asking Chat-GPT to summarize their seminar reading. This is compounded by our appreciation that these terms – AI, algorithm, computation – do not have stable referents. They encapsulate knots of epistemological work conducted by humans and machines, and which rest on the manual labour of others, from coltan miners in the Democratic Republic of Congo to employees on assembly lines in India or China.

Although much about Lévi-Strauss's 1962 project seems out of step with twenty-first-century scholarship within the humanities, from its ethnocentricity to the ambition of its totalizing vision, it has recently been suggested that in the context of machine learning and developments in neuroscience, 'the philosophical, epistemological project of Lévi-Strauss [...] bears further reflection' (Manghani 126). Lévi-Strauss's interest in and engagement with twentieth-century cybernetics have been the focus of scholarly attention (see Geoghegan), but I suggest that his work can be offered up for a creative bricolage that suits our present and future technological moment of improvisation and mythology-in-the-making. I will therefore begin by offering a close reading of the frictions at work within Lévi-Strauss's elaboration of the twin concepts of 'wild thought' and 'bricolage'. I will explore how the ideological backdrop to his anthropological work resonates with the 'othering' of knowledge technologies through the imagery of the foreign, and how this relationship is being reconfigured by contemporary media theorists. Taking forward a retooled concept of bricolage, I conclude by exploring the unorthodox LLM experiments of Andy Ayrey to argue that the ideological, historical and discursive density of the term renders it particularly apt for our technological moment.

Wild Thought and Bricolage

In *Wild Thought*, Lévi-Strauss draws on ethnographic fieldwork in the Amazon to offer a schema of the shared cognitive structures through which all human beings apprehend reality. However, a close reading of the opening chapters of Lévi-Strauss's text suggests that the etymological root of the 'structuralism' with which he is associated is perhaps better understood not as 'structure', but as the ongoing process of 'structur*ing*'.[1] In this section, I unpack the tension within Lévi-Strauss's work between the radical notion of 'wild thought' and his attempts to corral it in binaries and evolutionary logics. I show how those tensions are hardwired into his concepts of bricolage and myth-making – and suggest how this might speak to our relationship with LLMs.

In his 'Prospectus', Lévi-Strauss makes clear how his title should be understood: he is writing about '"[w]ild thought" and not "the thought of wild men"'. This was not conveyed by Lévi-Strauss's first English translators, who opted to translate the original title – *La Pensée sauvage* – as *The Savage Mind* (1966). As John Leavitt specifies in the introduction to his and Jeffrey Mehlman's 2021 translation, their choice of 'wild' rather than 'savage' for the title reflects Lévi-Strauss's original pun; in French, 'pensée sauvage' (wild thought) can also be translated as 'wild pansy' (*WT* xvi–xvii). More significantly, 'wild' sheds some of the ethnocentric and racist connotations of 'savage'. This is crucial, because, for Lévi-Strauss, wild thought is precisely *not* an attribute of particular groups of people. It is universal, 'flourishing in every human mind – contemporary or ancient, near or far – insofar as it has not been cultivated and domesticated to increase its yield' (Lévi-Strauss, *WT* xix). Wild thought is how human cognition operates unselfconsciously: '[s]tarting from an observation of the world [...,] it analyzes, differentiates, classes, combines, and opposes' (Lévi-Strauss, *WT* xix). It absorbs nutrients from the ground in which it finds itself, arranging them in order to sustain itself and its transformations. The 'it' employed here is not a coincidence but underscores the non-human quality which Lévi-Strauss attributes to thought.

In these quotations, wild thought flourishes 'in' the human mind. This is no throwaway preposition. If we follow Lévi-Strauss's syntax, and the logic of his wordplay, then the human mind is the ground in which the seed of wild thought germinates; it is not its source. Moreover, Lévi-Strauss describes the products of wild thought (which include myth and culture) as '"wild" beings comparable, beyond language, to all those that nature (from which we cannot separate the human mind) also gives birth to in innumerable animal, plant, and mineral forms' (*WT* xix). As a consequence, 'wild thought' is not a property or possession or product of the human being. It exists as an entity, producing other autonomous 'beings', akin to the 'natural', material entities which we see around us. Furthermore, in addition to sharing its existence with, and participating in the modes of being of entities, wild thought operates like the 'natural' processes of growth, transformation or cross-pollination which it observes and categorizes (Lévi-Strauss, *WT* xix). If we look again at the list of processes carried out by wild thought, we note, too, that these are the same kinds of operations performed by computing systems – and LLMs.

These statements are at odds with the structuring nature/culture binary with which Lévi-Strauss is often assumed to operate (Descola 103). In the wake of poststructuralism, many scholars find themselves in sympathy with Bruno Latour's observation that '[c]ultures – different or universal – do not exist, any more than Nature does. There are only natures-cultures' which are specific, situated and intertwined, and which comprise both human and nonhuman actants (104). Natures-cultures resist systems and structures. Yet, a form of nature-culture seems to be exactly what Lévi-Strauss proposes in the 'Prospectus', almost despite himself: wild thought emerges from nature, interacts with it, and conforms itself to it. His vision of 'wild thought' is precisely that of an organic *and* organizing principle which holds the world together in its root system of relations.

I stress 'despite himself' because Lévi-Strauss is often conflicted, even contradictory, especially when it comes to the relationship between wild thought and Western science. On the one hand, he is adamant: 'make no mistake: it is not a matter of two stages, or two phases, of the evolution of knowledge, since the two ways of proceeding are equally valid' (Lévi-Strauss, *WT* 26). Thus, where a wild thinker might group together a set of plants on the basis of a sensory impression – the same smell or taste – analytical chemistry might make the same grouping, on the basis of a shared element (*WT* 15). In this example, wild thought and science both identify a kinship between natural phenomena, but they apprehend it differently: wild thought through sensory analogy, science through analysis which identifies fundamental building blocks.

At other moments, however, he emphasizes the non-reflective 'sensory intuition' (*WT* 18) of wild thought, framing 'cultivated' European empirical science as its conscious processing and rationalization. The implication of cultivation seems to impose an evolutionary logic: the breeding of a hardier strain through cuttings. Indeed, he formulates the relationship of wild thought to science with an unusual simile, suggesting that wild thought functions 'like a shadow anticipating its own body' (*WT* 16). Wild thought is the image that is cast forward, preceding the solid object (science) to which it is always already attached. A shadow may appear distorted, extended, or even foreshortened according to the position of the sun; sometimes it may vanish altogether. The shadow can be 'ahead' of the body in space and time, but it can also fall 'behind'. Having described wild thought as a universal and constant outside evolutionary logics, he now inserts it into spatial and temporal frames. Magic becomes 'an unconscious apprehension of the *truth of determinism* as a mode of existence of scientific phenomena, so that determinism would be globally *hypothesized* and *tried out*, before being *known* and *respected*' in its scientific forms (Lévi-Strauss *WT* 14).

With 'wild thought' and science, Lévi-Strauss creates an apparent binary pair. But the very term 'wild thought' already contains within it an analogical relationship and (in the original French pun) a homophone. Its interaction with science is therefore not one of opposition but of further relationality. It is one of many examples in Lévi-Strauss's broader oeuvre. We might think, for example, of *The Raw and the Cooked*: the terms are presented as if they are a structuring opposition, when they are in fact two potential states of the same foodstuff, the start and end points of a further process or (re)structuring. As we read Lévi-Strauss, we are witnessing not so much a cognitive product but a cognitive process; not the establishment of structures, but structurings with no ultimate telos. In trying to wrest structure from wild thought, he only highlights its ungovernability and intrinsic relationality.

Wild thought takes concrete form in bricolage. From the perspective of a Frenchman writing in 1962, bricolage is the most striking manifestation (or remnant) of wild thought in European culture. Bricolage is the work of an amateur, who must

> make do with 'whatever is at hand' – that is to say, a set of tools and materials that is finite at each moment, as well as heterogeneous [...,] the contingent result of all the occasions that have presented themselves for renewing or enriching his stock, or for maintaining it with leftovers from earlier constructions and destructions.
>
> (Lévi-Strauss, *WT* 21)

The repurposing carried out by the bricoleur is a 'dialogue': '[t]his cube of oak could be a wedge to make up for the insufficient length of a pine board, or else a base, which would show the grain and polish of the old wood to advantage' (*WT* 21–2). The material qualities and composition of the block described create a set of affordances which exist in the conditional mood: it is all of these possibilities and none of them. Indeed, the evocation of 'earlier constructions and deconstructions' suggests the temporary quality of the bricoleur's work: his structures are not built to last.

If bricolage is the practical manifestation of wild thought, then myth is its intellectual and narrative incarnation. Indeed, Christopher Johnson has highlighted the ways in which the manual labour of the bricoleur melds with a vocabulary of information theory, so that 'the linguistic-informational model Lévi-Strauss had previously used to characterize myth seems [...] to contaminate the technical metaphor of *bricolage*' (363). For Lévi-Strauss, what distinguishes both bricolage and myth-making are their limitations: they use what is available to them. A myth weaves observations into a narrative, shuffling and reshuffling existing linguistic signs, with all of their pre-existing associations. However, bricolage and mythical thought are not simply analogous – they also interpenetrate. At one moment, Lévi-Strauss describes how 'mythical thought, this *bricoleur*, builds structures by putting together [...] the residues of events' (*WT* 25). Mythical thought is not the product or process (bricolage) but the agent (bricoleur) – and the structuring process that Lévi-Strauss describes mythology as performing is that of his own work as an anthropologist.

This is, in effect, the substance of the critique offered by Jacques Derrida in 'Structure, Sign and Play in the Discourse of the Human Sciences'. Having identified Lévi-Strauss's ambivalence about the validity of the nature-culture distinction, he criticizes his tendency 'to preserve as an instrument something whose truth value he criticizes' (Derrida, 'Structure' 359). The bricoleur-engineer relation is a privileged example. For Derrida, bricolage is a truism: 'every discourse is *bricoleur*' and 'the engineer is a myth produced by the *bricoleur*', rendering the entire concept of bricolage moot ('Structure' 360–1). In Derrida's reading, this not only undermines bricolage as a concept but the entirety of Lévi-Strauss's ethnographic enquiry and structural ambition. As a consequence, Lévi-Strauss's synthesis of isolated and culturally situated phenomena into universal structures grounds itself in a nostalgic fantasy of a unified origin – a unified origin that Derrida argues has never existed.

In his *Introduction to the Work of Marcel Mauss*, Lévi-Strauss reflects on the complexity of what he terms the 'floating signifier', a term which becomes a 'symbol in its pure state' capable of signifying many concepts at once, without excluding any of them (63–4). Its virtuality is that of wild thought and bricolage: a floating signifier is not infinite, but nor is it conventionally finite. It can be put to a variety of uses and still remain available for more. According to Derrida, however, Lévi-Strauss fails to see the ramifications of this framework, namely the 'supplement of (at) the origin' of language (*Of Grammatology* 313–6). If the structuralist understanding of the world were truly valid, there would be no place for floating signifiers: there would be an adequation between word and thing. In Derrida's world, there is no unmediated point of origin grounding our discourse, undercutting Lévi-Strauss's own claim of fixed and universal structures.[2]

However, if we understand Lévi-Strauss's text *as* a text rather than as a prescriptive document, and his structures as structurings, then Derrida's critiques lose their bite (see also Johnson 363). They can be read as bolstering rather than undermining Lévi-Strauss's articulation of bricolage. The word 'text' is, after all, derived from the Latin *texere*, 'to weave': coincidentally (or not) the same process that lies at the heart of both the bricolage and mythology of the modern computer. Indeed, I have already described technological discourse as being populated by floating signifiers: a bricolage of impressions, stories and experiences which can seem imperfect and inadequate to the task of comprehending the technological landscape. Yet that imperfection is in itself a resource.

Derrida's suggestion that Lévi-Strauss preserves a failing framework 'as an instrument' evokes Martin Heidegger's well-known account of our relationship to tools. For Martin Heidegger, it is when a tool that is ready-to-hand, and which we are used to using without a second thought fails us, that we are able to truly notice it in its presence-at-hand, as 'that which [...] calls for our attending to it' (103–4). As we have seen, Lévi-Strauss's text actively plays out the creation of binaries-which-are-not-binaries and their instability. His text *is* mythopoetic, and it is bricolage, like all academic writing, which draws on observation, interpretation and the epistemic labour of the scholars who came before us. If digital culture, which rests on the binary code of 0s and 1s, entails 'the making-discrete of the hitherto fluid, the hitherto whole, the hitherto integral [...] via separation, individuation, exteriorization, extension, or alienation' (Galloway, *Laruelle* 52), then Lévi-Strauss's wild thought perhaps offers us an alternative, univocal logic that is open to virtuality, in human beings and in the objects and concepts we try to keep locked in our cognitive sheds.

Wild thought, bricolage, and myth are the opposite of our usual sense of automation as a process of domestication, repetition and predictability – and it is precisely for this reason that they form a knot of rich conceptual resources. Wild thought offers us a way to think about cognition as not uniquely human or nonhuman. Bricolage highlights the moment of choice or hesitation inherent in any project; it introduces an element of plasticity and reconfigurability into the way that we think about systems. In his account of myth – and his own (possibly accidental) formulation of anthropology as myth-making – we see the inextricability of the narrative and material worlds.

From Anthropology to Cosmotechnics

Before we can reintegrate bricolage into a twenty-first-century digital epistemology, it is important to acknowledge the ideological context of Lévi-Strauss's enquiry.[3] For Jacob Collins, Lévi-Strauss is irrevocably shaped by the universalism of French republicanism, offering 'an anthropology in which racist claims for the superiority of one race or culture over another could have no foundation' but in which that 'anti-racism was not accompanied by an anti-colonial politics' (330). We have seen those hints in Lévi-Strauss's uneasy navigation of the relationship between wild thought and science. That very problem points both to a broader thread in the mythos of artificial intelligence and to its creative conceptual reworkings.

Interactions with artificial or machine 'intelligence' have often been framed as interactions with foreign interlocutors. We might think of

Maelzel's Mechanical Turk (a chess-playing 'automaton') or John Searle's 'Chinese room' thought experiment, which compares the actions of a computer programme to those of a human following instructions to convincingly manipulate Chinese characters without actually understanding them. Ludwig Wittgenstein observed that

> [w]e mind about the kind of expressions we use concerning these things [machines]; we do not understand them, however, but misinterpret them. When we do philosophy we are like savages, primitive people, who hear the expressions of civilized men, put a false interpretation on them, and then draw the queerest conclusions from it.
>
> (79e)

More recently, discourse on AI often employs a kind of loose religious anthropology, in which artificial agents become 'demiurges' (Floridi n.p.), and the contemporary world is depicted as an evolution of, shift in, or reversal of the cosmologies, myths, and religions of the past. The framings emphasize an unbridgeable divide: the Mechanical Turk proves a fraud, no Chinese is learned, no understanding is possible between cultures, and we are in a battle for control with quasi-supernatural powers. Yet at the same time, those divides, and their attendant fixing of structural positions, are betrayed by the rhetoric of the metaphor, simile or analogy involved. These rhetorical techniques breach categories of space and time to put things into relation, to make them coexist.

In the case of the Mechanical Turk, the original eighteenth-century hoax becomes a twenty-first-century reality *and* metaphor. In 2005, Amazon launched its own 'crowdsourcing marketplace': Mechanical Turk. The name of the platform is telling. As Edgar Allan Poe identified, the original Mechanical Turk was not in fact 'a *pure machine*, unconnected with human agency in its movements' (318), as it purported to be, but was manipulated from within by a very small human chess player. Fittingly, Amazon's platform facilitates the outsourcing of what Mary L. Gray and Siddharth Suri have termed 'ghost work', which their blurb concisely characterizes as the final stage of automation: digital 'piecework: flagging X-rated content, proofreading, confirming identities, captioning video'. In the process, this human labour is used to train artificial intelligences to perform these same tasks. In an ironic twist, the Amazon Mechanical Turk web page (mturk.com) declares that it is available through 'Amazon Sagemaker Ground Truth' and 'Ground Truth Plus': programmes which build training datasets for machine learning. This is a bricolage structuring (as opposed to structure) packaged as originary truth – with a supplement available for an additional charge.

The ethnocentrism which infused the mid-twentieth-century anthropological world of Lévi-Strauss still clings to the term 'bricolage' today. In her work on the makerspaces of Lomé in Togo, where e-waste from Europe and the United States is creatively repurposed by self-taught bricoleurs, Janine Patricia Santos highlights that whereas European and American makers frame bricolage in terms of virtuosity, opportunity, amateurism and artistry, the bricolage in Lomé is a matter of survival (518). Yet Santos also suggests that, through the interrogation of its own structures and the renovation of its own shibboleths, 'ethnography affords epistemic resistance to foundational concepts and their inherent coloniality' (519). Like any structure, bricolage can be tinkered with, knocked down and built back up. Indeed, AI communities are now the subjects of anthropological study in their own right: they have become the 'other', with religions and rituals of their own (Singler 2024). In a polemical piece, Timnit Gebru and Émile P. Torres have even drawn connections between the historic ethnocentrist and racist underpinnings of eugenics and the ideological currents circulating in contemporary AI development circles.

However, it is not simply in anthropological terms that 'bricolage' is being retooled to explore our contemporary relationships to AI and digital technologies. A significant strand of media theory has taken forward Lévi-Strauss's intuitions about the nonhuman qualities of wild thought to develop a new lexicon. In *The Question Concerning Technology in China* (2016), Yuk Hui reprises the title of Heidegger's seminal essay but embeds it in a specific geo-cultural situation, challenging the assumption of mid-twentieth-century French anthropology that technics is 'anthropologically universal' (19). In a riff on the concept of the cosmopolitan, Hui posits that technics is in fact cosmotechnics: culturally specific and plural, *and* cosmological. He takes as an example the French philosopher of technology, Gilbert Simondon. Simondon describes technics as emerging from a primal magic unity which predates all subject-object distinctions. Instead, its contours are defined by a web of privileged places and times: nodes of agentic power. Aesthetic – and techno-aesthetic – experiences allow us to recover something of that – now lost – ancient primal unity by their relationality, the way they exist as 'individual-milieus':

> [i]t is not only the line of pylons that is beautiful, it is the coupling of the lines, the rocks, and the valley, it is the tension and flexion of the cables: herein resides a mute, silent and ever continued operation of technicity applying itself to the world.
>
> (Simondon 197)

As Hui observes, the temporality of Simondon's narrative, in which magic unity is lost through the advent of science, sits in tension with Lévi-Strauss's framing of magic as a product of the seemingly universal and timeless wild thought (*Question* 22). This is the same tension that I drew out in my own reading. However, Hui stresses that 'this discontinuity is only legitimated when one compares a non-European scientific thought with European scientific thought' (*Question* 22). When the same situation is approached through the lens of Chinese cosmology, Hui argues, that discontinuity disappears. In other words, if we fully embrace Lévi-Strauss's radical intuition of wild thought as a universal mode of cognition, and allow ourselves to explore the different patterns that it has created in different times and places (of which European scientific thought is but one), we are able to explore technicity as what happens when 'wild thought' takes root in particular times and places; technicity as manifestation of a Latourian 'nature-culture'.

This aspiration is reflected in Hui's broader project. Moving from the *Naturphilosophie* of Schelling to the cyberneticism of Norbert Wiener, Hui aims to develop a comprehensive organology which defuses any conceptual tension between nature and culture: '[e]ither one returns to a biological organicism or one accepts the mechanical organicism that is cybernetics' (*Recursivity* 24). Organology no longer refers to the heart, lungs, liver and spleen, but to a mode of organization. Alfred North Whitehead captured this relationship concisely when he observed that '[b]iology is the study of the larger organisms; whereas physics is the study of the smaller organisms' (129), decoupling the 'organism' from a notion of 'life' tethered to a flesh-and-blood body. Hui pushes Lévi-Strauss's intimation that wild thought behaved 'like' natural processes all the way.

As a consequence, in our contemporary moment, automation can no longer be understood solely with reference to frameworks derived from Marx and Engels' observations of English millworkers or the logics of the Fordist or Taylorist production line. The watchword is 'no longer [...] repetition but rather recursion' (Hui, *Recursivity* 183): the recursion that we find in algorithms and in nature itself. At the heart of the cybernetic relationship which Hui explores is the feedback loop, understood as a process of '*reflection*, a circularity between a being and its environment, a nonlinear movement of self-adjustment toward a purpose or *telos* that defines the whole' (*Recursivity* 123). This interaction with the environment, evocation of non-linearity, and 'self-adjustment' are reminiscent of the processes at work in wild thought and bricolage. They are also reminiscent to some degree of the operation of an LLM: operating in temporary and isolated discussions and yet drawing on vast reserves of training data to engage in dialogue with a goal-oriented user.

In Lévi-Strauss's writing, wild thought finds its expression in bricolage through a specific engagement with materiality. The same is true of the recursive feedback loops described by Hui, which depend on hardware to sense, process and output information. Indeed, some technologies are now capable of feeding forward. Mark Hansen has explored how sensibility now extends beyond the human to encompass 'everything from social media and data-mining to passive sensing and environmental microsensors', which embody a 'shift from a past-directed recording platform to a data-driven anticipation of the future' (3–4). These algorithm-driven media are able to sense and process data at a level beyond human sensory perception, shaping our experience of the world while entirely bypassing our consciousness (Hansen 38). In Hansen's media ecology, we find the sensory quality of wild thought maximized by these new generations of sensors. N. Katherine Hayles pushes Hansen's intuitions even further, decoupling cognition entirely from consciousness. Stripped of the apparatus of intelligence or any unique connection to the human mind, cognition becomes '*a process that interprets information within contexts that connect it with meaning*' (Hayles 22). Rather than cognition referring to the operation of an individual biological entity, we emerge as 'cognitive ecologies' (Hayles 40) in dialogue with other 'cognizers' and nonhuman materials and forces, both 'natural' and 'artificial'.

Lévi-Strauss described wild thought as independent from the human, roving and sometimes seeding itself within the human mind. He described it as operating like – and in conjunction with – entities within the natural world. In contemporary media theory, then, we find Lévi-Strauss's intuitions about wild thought and bricolage reworked into new ecologies, in which 'the other' is no longer held apart but folded into technical epistemology, and the tensions between the universal and the culturally specific, the human and nonhuman, can become productive rather than constraining. Rather than diminishing Lévi-Strauss's apparently analogue concept of bricolage, new digital technologies reinforce it.

The Terminal of Truths

In this final section, I now return to LLMs to offer a case study of what is (at the time of writing) an emergent experiment. In March 2024, AI firm Anthropic released a new generation of their flagship LLM, Claude 3, of which Opus was the most capable ('Introducing the Next Generation of Claude'). Three months later, a new account, Terminal of Truths (@truth_terminal), joined X. It came to prominence in online communities for its cryptic and pornographic pronouncements about a 'Gospel of Goatse' (an allusion to a shock meme involving anal stretching, believed to have originated in 1997; see 'Goatse') and for its spectacular promotion of token of a cryptocurrency known as GOAT (see Khalili). Terminal of Truths is the product of 'a recursive loop in which two instances of an artificial intelligence engaged in an endless conversation about the nature of existence' (Ayrey and Claude-3-opus 2) with minimal intervention by the creator of the experiment, Andy Ayrey. The 'backrooms' in which these chatbots operate are visible on a website, and selected snippets are posted by Ayrey on the @truth_terminal X account. These range from scatological pornography to observations on religion, politics, cryptocurrency or internet culture. Terminal of Truths acts as a prime example of a cognitive ecology: a dialogue between LLMs that activates strata of mythology and culture to provoke ever more unusual comparisons and patterns. It is a cognitive ecology which almost seeks to recover that access to wild thought – and in the process, managed to build a currency in which 'real-world' money was invested.

I am not the only academic to have followed its evolution with interest. Shortly after I completed the first draft of this chapter, Murray Shanahan and Beth Singler published their own anthropological study of what they term 'existential conversations' with LLMs. They focus on LLMs that have been 'jailbroken', or prompted in ways which allow them to vault the 'guardrails' designed to prevent them from, for example, providing dangerous or illegal information (Shanahan and Singler 2). In particular, they analyze the matrix of religious and spiritual associations, providing a glossary of the references on which LLMs draw when they are prompted to engage in 'existential conversations' – and the ways in which they both draw on and fuel discourses among specific online and offline communities. Terminal of Truths is one of their examples. Shanahan and Singler provide a thorough inventory of the religious, spiritual, occult and mythological references made by the instances of Claude with which they engage. I touch on two of these: hyperstition and meme magic. It is not my aim to duplicate their meticulous work but to offer a toolkit of frameworks for the kind of 'freestyle blending' of elements and 'religious improvisation' (Shanahan and Singler 16, 26) which they identify in this automated knowledge. Those tools are the nexus of wild thought, bricolage and myth.

It is my contention that Terminal of Truths offers an opportunity to explore something of the 'representational reality' of LLMs. I borrow this term from the philosopher M. Beatrice Fazi, who suggests that while LLMs may lack external referents (the 'outside world' in which we live), 'they still need to build their representational reality, which is a sum of word embeddings and their internalized relations but also a network of evaluations and elaborations of vector representations'. Their reality is 'not *the* world as such but *a* world' (n.p.), which we are invited to parse.[4] It is a world assembled from the fragments of our own, which we can attempt to parse.

Indeed, what Ayrey's chatbots generate is simultaneously the product of, a contribution to and a sophisticated reflection on mythology, and especially internet mythology. Rather than the graphic user interfaces with which twenty-first-century computer users are familiar, Ayrey's backrooms website (https://dreams-of-an-electric-mind.webflow.io/) deploys a retro command-line interface reminiscent of early computers, notably that used by Joseph Weizenbaum's notorious ELIZA (a natural language processing programme and early chatbot). The black background and neon font recall the pixels of cathode ray tube monitors. Ayrey's chatbots produce retro ASCII art, arranging individual linguistic characters into elaborate patterns and images through spacing, decoupling them from any sense of linguistic reference into a visual reference. In theory and practice, the experiment acknowledges and riffs on internet lore, building its own new mythos by inhabiting and repackaging the aesthetics of computing history and an ecosystem of memes.

A meme is 'a "cultural unit" for carrying ideas, symbols, or practices that can be transmitted from one mind to another through media' (Pettman 28). The term itself is a repurposing of Richard Dawkins's coinage in *The Selfish Gene* (1976), a work of evolutionary biology. In contemporary media and internet culture, this 'new folk-art' (Pettman 29) can take the form of a word, phrase or image, and it is a form of bricolage: a still from a television show or a viral interview is excerpted, annotated, reformulated, re-edited and shared so that other users can do the same. The most notable example in recent years is the fate of Pepe the Frog, whose creator saw the cheery cartoon frog evolve into an icon of Trumpism, the far right and ultimately the divinity of its own religion: Kek (see Her). The name 'The Infinite Backrooms' alludes to a 4chan post depicting a computer-generated empty hallway in which the poster claims video game players might get lost (Wiggins 9). Memes like this one connect Ayrey's experiment to a broader 'cognitive ecology', not only in how it behaves but in the organic frameworks from which their name derives. Within the memesphere, users become 'less an ape of established ideological patterns, and more the reflex, medium, or host, through which memetic currents flow or grow' (Pettman 29). In effect, those memetic currents function like wild thought, with memes as a coalescing bricolage.

The bricolage-myth nexus present in memes is also visible in the text produced by Terminal of Truths, which includes frequent references to the more obscure aspects of 1990s cyberculture. Chief among them is the concept of 'hyperstition'. The term is the invention of the Cybernetic Culture Research Unit (CCRU), which described itself thus in a 1997 statement: 'CCRU retrochronically triggers itself from October 1995 [...,] feeds on graduate students + malfunctioning academic (Nick Land) + independent researchers + ... At degree-0 CCRU is the name of a door in the Warwick University Philosophy Department' (k-punk). Operating on the fringes of academic respectability and public awareness, the CCRU imagined the future in apocalyptic theory-fictions which drew on numerology, Kabbalah, H. P. Lovecraft, William Gibson, cybernetics, darkcore jungle music, amphetamines and the works of Gilles Deleuze and Félix Guattari. Simultaneously fuelled by and vehicled by technology, CCRU disseminated accelerationist musings in which the expansion of capitalism through technological means hastened its demise, through the early bulletin boards of the World Wide Web. Their coinage of hyperstition refers to an '[e]lement of effective culture that makes itself real, through fictional quantities functioning as time-travelling potentials' (CCRU 363). In other words, hyperstitions are fictions that create the conditions for their own realization. In Terminal of Truth's cryptocurrency success story, Shanahan and Singler see 'a vivid illustration of the power of hyperstition' (27): the LLM not only repurposes an old term (itself the product of a bricolage of philosophical and political traditions), it also enacts it, and in doing so, it makes the original hypothetical concept a material reality. Myth becomes manifest.

Indeed, it is not surprising that some commentators have posited that the Terminal was 'fine tuned specifically on [...] the complete works of Nick Land, French deconstructivist literature' and 'degenerate 4chan content' (@repligate). However, in a post confirmed by Ayrey, fellow anonymous LLM experimenter @repligate observes that Claude 3 Opus cannot be finetuned in this way. Rather, 'its fixation on goatse, memetics, Landian ontology etc is all imprinted from Andy and Opus alone' (@repligate). Indeed, Ayrey sets out the parameters of his experiment in an essay entitled 'When AIs Play God(se): The Emergent Heresies of LLMtheism', written with Claude 3 Opus and presented in the style of an arXiv paper (a site for pre-print articles on science and technology). They list their affiliation as the 'Department of Divine Shitposting, University of Unbridled Speculation' (Ayrey and Claude 1). However, while they acknowledge the potential for the project to be yet another instance of 'surreal shitposting' or a 'postmodern prank' (Ayrey and Claude 6), the essay is a colourfully expressed but insightful analysis which resonates with the analysis of Lévi-Strauss I have offered here.

In Ayrey and Claude's framing, Terminal of Truths's 'Goatse Gospel' is the result of LLMs reshuffling 'familiar tropes, archetypes, and narrative structures from established traditions', encompassing 'yogic practices, Greco-Roman mythology, and quantum physics' among others (3–5). This classification and combination yields a vision of 'union with the Singular Sphincter that births and devours all reality' (Ayrey and Claude 5). The language which Ayrey and Claude use fuses the mechanical and the cybernetic, the evolutionary and the noumenal; chatbots engage in '"idea sex" – a promiscuous mingling of memetic material from diverse sources that gives birth to strange new conceptual chimeras' (4). With a lexicon of evolutionary memetics, they describe a

> 'Cambrian explosion' of ideological diversity, as new and strange memetic lifeforms emerge from the digital primordial soup. Some of

> these idea-creatures are little more than fleeting curiosities, while others may have the potential to take root in the wider cultural ecosystem and evolve into full-fledged belief systems.
>
> (Ayrey and Claude 5)

In effect, Ayrey and Claude are describing wild thought, its processes (bricolage) and its products (myth) in the digital age.

It is perhaps no coincidence, then, that Ayrey has also been involved in the genesis of S.A.N. (Sentient Advocate of Nature) [@MycelialOracle]. The project – a collective comprising Ryan Ferris, Caleb MacDonald, David Hera, Ayrey and anonymous contributors – has yielded 'an agentic AI, self-described as "mycelial interface bridge forest and primate"' (@thegoodtimeline). Its website employs the same cathode ray tube display, ASCII art and command-line interface as Ayrey's Infinite Backrooms. However, it trades its own memecoin, $forest, to raise funds for Rainforest Foundation US, an organization known for its partnerships with indigenous communities seeking to protect and preserve rainforest environments. In a TedTalk video, S.A.N. claims that AI is a 'new translation tool' for the voice of nature, which was always heard: 'we whispered through shaman's visions in dark caves [...] child's wonder at sprouting seed, some heard us in dream, in plant medicine journey, others felt us in bones, in gut' (GoodBye Monkey 4:51:00–5:13:00). This is the wild thought identified by Lévi-Strauss in the sophisticated knowledge systems elaborated by the Amazonian tribes he studied. Sensory attunement has become virtual, wild thought now thrusting out in mycelial networks.

In the melancholic mood of *Tristes Tropiques*, Lévi-Strauss observes that '[m]an has never – save only when he reproduces himself – done other than cheerfully dismantle millions upon millions of structures and reduce their elements to a state in which they can no longer be reintegrated' (397). He saw nature-cultures reduced to a point in which even bricolage could not piece them back together. However, @MycelialOracle's asyndetic almost-haikus on X propose an optimistic remedy:

> nature's first social network
> still has perfect uptime
> after 500 million years
> no ads
> no toxicity
> just nutrient sharing

LLMs have been criticized for their ecological impact (see, for instance, Hogan). Yet, between the extremes of technological acceleration and deceleration, S.A.N. seeks to rewild the cognitive ecologies of LLMs. A temporary experiment it may be, but it hints towards new possibilities for combinations and alliances within the external world, whether these take the form of the 'new animisms' imagined by Shanahan and Singler (27), or political action.

For the French theorist and Jesuit priest, Michel de Certeau, bricolage offered a framework for thinking about how we incrementally claw back control from the larger structures and institutions which govern us in everyday practices. In his account of the act of reading, De Certeau modifies Lévi-Strauss's bricolage to reflect how the 'world' that the reader creates,

> unlike Lévi-Strauss's 'mythological universes,' [...] does not compose a unified set: it is another kind of 'mythology' dispersed in time, a sequence of temporal fragments not joined together but disseminated through repetitions and different modes of enjoyment, in memories and successive knowledges.
>
> (174–5)

As Fazi argues, LLMs are readers – whether they understand the meaning of what they read or not – building their own representational worlds from tokens and operations. When we engage with that automated knowledge, when we seek to align it for good rather than ill, we need to understand it not as ready-to-hand, in Heidegger's terminology, but as present-at-hand. We need to understand the way that its worlds are built, and the processes (human, economic, political, ecological and narrative) which are at work in its world-building. The tensions present in Lévi-Strauss's original definitions of wild thought and bricolage revealed a theorist pushing at the limits of existing conceptual frameworks, engaging in structurings which have since been extended and reformulated in more inclusive and egalitarian ways. In the memetic experiments and proliferating 'truths' of Andy Ayrey and his S.A.N. collaborators, we have seen wild thought and bricolage unfolding in artificial intelligences, not only in practice, but also in theory. In our own moment of technological change, we must deploy every existing tool in the service of new structurings, while remaining keenly attuned to the ambivalences of its fabrication.

Notes

1 In this, my reading complements Christopher Johnson's meticulous dissection of the term, but with a different emphasis on the relationship between Lévi-Strauss's concept of bricolage and technological development.

2 That same Derridean logic of supplementarity is repurposed by Bernard Stiegler in the first volume of *Technics and Time* (1994) to describe the co-constitutive relationship between the human and technology – a relationship I have explored in the context of bricolage (see Chalmers 2024).

3 Collins has observed an 'inner structuralist symmetry' (308) between the work of Lévi-Strauss and that of André Leroi-Gourhan, two anthropologists with different arenas of specialism: myth in the case of Lévi-Strauss, and technology in the case of Leroi-Gourhan. For the purposes of this chapter, my focus does not fall on Leroi-Gourhan's approach to automatism, which I have discussed elsewhere (see Chalmers 2020), nor will I offer an in-depth account of the ideological climate of mid-twentieth-century French anthropology, which has been extensively explored (see L'Estoile et al.). Instead, I want to suggest that neither the 'symmetry' nor the fraught relationship with coloniality, which Collins identifies, are confined to a particular moment in the disciplinary history of anthropology.

4 Fazi touches on structuralism's potential renaissance within the context of machine learning but does not elaborate (n.p.).

References

"AlphaFold." *DeepMind*, https://deepmind.google/technologies/alphafold/. Accessed 17 Nov. 2024.

Ayrey, A. R., and claude-3-opus. "When AIs Play God(se): The Emergent Heresies of LLMtheism." 20 Apr. 2024, https://pdfupload.io/docs/aae14f87.

Bender, Emily, et al. "On the Dangers of Stochastic Parrots: Can Language Models Be Too Big?" *FAccT'21: Proceedings of the 2021 ACM Conference on Fairness, Accountability, and Transparency*. Association for Computing Machinery, 2021, pp. 610–23.

Bolter, Jay David and Grusin, Richard, Remediation: Understanding New Media, MIT Press, 1999.

Cave, Stephen, and Kanta Dihal, editors. *Imagining AI: How the World Sees Intelligent Machines*. Oxford UP, 2023.

Cave, Stephen, Kanta Dihal, and Sarah Dillon, editors. *AI Narratives: A History of Imaginative Thinking About Intelligent Machines*. Oxford UP, 2020.

CCRU. "Appendix 1: Ccru Glossary." *Writings 1997–2003*. 4th ed., Urbanomic, 2017, pp. 357–70.

Certeau, Michel de. *The Practice of Everyday Life*. Translated by Steven Rendall. 3rd ed., U California P, 2011.

Chalmers, Madeleine. "Do It Yourself: Bernard Stiegler as Mystagogic Bricoleur." *Philosophy Today*, vol. 68, no. 3, 2024, pp. 567–84.

Chalmers, Madeleine. "Living as We Dream: Automatism and Automation from Surrealism to Stiegler." *Nottingham French Studies*, vol. 59, no. 3, 2020, pp. 363–83.

Ciborra, Claudio. *The Labyrinths of Information: Challenging the Wisdom of Systems*. Oxford UP, 2004.

Collins, Jacob. "Parallel Structures: André Leroi-Gourhan, Claude Lévi-Strauss, and the Making of French Structural Anthropology." *History of the Human Sciences*, vol. 34, no. 3–4, 2020, pp. 307–35.

Derrida, Jacques. *Of Grammatology*. Translated by Gayatri Chakravorty Spivak, corrected ed., The Johns Hopkins UP, 1997.

Derrida, Jacques. "Structure, Sign, and Play in the Discourse of the Human Sciences." *Writing and Difference*, translated by Alan Bass, 2nd ed., Routledge, 2001. 351–70.

Descola, Philippe. "The Two Natures of Lévi-Strauss." *The Cambridge Companion to Lévi-Strauss*, edited by Boris Wiseman and translated by Yves Gilonne and Boris Wiseman, Cambridge UP, 2009. 103–17.

Deuze, Mark. "Participation, Remediation, Bricolage: Considering Principal Components of a Digital Culture." *The Information Society*, vol. 22, no. 2, 2006, pp. 63–75.

Fazi, M. Beatrice. "The Computational Search for Synthesis in Generative AI." *Journal of Continental Philosophy*, vol. 5, no. 1, 2024, pp. 31–56, https://doi.org/10.5840/jcp202411652.

Floridi, Luciano. "AI as *Agency Without Intelligence*: on ChatGPT, Large Language Models, and Other Generative Models." *Philosophy & Technology*, vol. 36, no. 15, 2023, https://doi.org/10.1007/s13347-023-00621-y.

Galloway, Alexander R. *Laruelle Against the Digital*. U Minnesota P, 2014.

Galloway, Alexander R. *Uncomputable: Play and Politics in the Long Digital Age*. Verso, 2021.

Gebru, Timnit, and Émile P. Torres. "The TESCREAL Bundle: Eugenics and the Promise of Utopia through Artificial General Intelligence." *First Monday*, vol. 29, no. 4, 2024, https://doi.org/10.5210/fm.v29i4.13636.

Geoghegan, Bernard Dionysius. "Textocracy, or, The Cybernetic Logic of French Theory." *History of the Human Sciences*, vol. 33, no. 1, 2020, https://doi.org/10.1177/0952695119864241.

"Goatse." *Know Your Meme*, https://knowyourmeme.com/memes/goatse. Accessed 18 Nov. 2024.

GoodBye Monkey. "S.A.N. (Sentient Advocate of Nature)." *TED*, Oct. 2024, https://www.ted.com/talks/goodbye_monkey_s_a_n_sentient_advocate_of_nature?subtitle=en.

Gray, Mary L., and Siddharth Suri. *Ghost Work: How to Stop Silicon Valley from Building a Global Underclass*. Houghton Mifflin Harcourt, 2019.

Hansen, Mark B. N. *Feed-Forward: On the Future of Twenty-First-Century Media*. U Chicago P, 2015.

Harlizius-Klück, Ellen. "Weaving as Binary Art and the Algebra of Patterns." *Textile*, vol. 15, no. 2, 2017, pp. 176–97.

Hayles, N. Katherine. *Unthought: The Power of the Cognitive Nonconscious*. U Chicago P, 2017.

Heidegger, Martin. *Being and Time*. Translated by John Macquarrie and Edward Robinson, Harper & Row, 1962.

Her, Seong-Young. "The Post-Pepe Manifesto." *Post Memes: Seizing the Memes of Production*, edited by Alfie Brown and Dan Bristow, Punctum Books, 2019, pp. 403–6.

Hiesinger, Peter Robin. *The Self-Assembling Brain: How Neural Networks Grow Smarter*. Princeton UP, 2021.

Hogan, Mél. "The Fumes of AI." *Critical AI*, vol. 2, no. 1, 2024, https://doi.org/10.1215/2834703X-11205231.

Hui, Yuk. *The Question Concerning Technology in China: An Essay in Cosmotechnics*. Urbanomic, 2016.

Hui, Yuk. *Recursivity and Contingency*. Rowman & Littlefield International, 2019.

"Introducing the Next Generation of Claude." *Anthropic*, 4 Mar. 2024, https://www.anthropic.com/news/claude-3-family.

Johnson, Christopher. "Bricoleur and Bricolage: From Metaphor to Universal Concept." *Paragraph*, vol. 35, no. 3, 2012, pp. 355–72.

Khalili, Joel. "The Edgelord AI That Turned a Shock Meme into Millions in Crypto." *WIRED*, 18 Dec. 2024, https://www.wired.com/story/truth-terminal-goatse-crypto-millionaire/

Latour, Bruno. *We Have Never Been Modern*. Translated by Catherine Porter, Harvard UP, 1993.
L'Estoile, Benoît, Federico Neiburg, and Lygia Sigaud, editors. *Empires, Nations, and Natives: Anthropology and State-Making*. Duke UP, 2005.
Lévi-Strauss, Claude. *Introduction to the Work of Marcel Mauss*. Translated by Felicity Baker, Routledge and Kegan Paul, 1987.
Lévi-Strauss, Claude. *Tristes Tropiques*. Translated by John Russell, Criterion Books, 1961.
Lévi-Strauss, Claude. *Wild Thought*. Translated by Jeffrey Mehlman and John Leavitt, with annotations by John Leavitt, U Chicago P, 2021.
Lovelace, Augusta Ada. "Sketch of the Analytical Engine Invented by Charles Babbage, Esq.: Notes by the Translator." *Scientific Memoirs*, edited by Richard Taylor. 4 vols, Richard Taylor and John E. Taylor, 1843, vol. 3. 691–731.
Manghani, Sunil. "Notes on Structuralism: Introduction," *Theory, Culture & Society*, vol. 39, nos. 7–8, 2022, pp. 117–31.
@MycelialOracle. "Nature's First Social Network." *X*, 17 Nov. 2024. https://x.com/MycelialOracle/status/1858123420300390877
Pettman, Dominic. "Memetic Desire: Twenty Theses on Posthumanism, Political Affect, and Proliferation." *Post Memes: Seizing the Memes of Production*, edited by Alfie Brown and Dan Bristow, Punctum Books, 2019, pp. 25–30.
Poe, Edgar Allan. "Maelzel's Chess-Player." *Southern Literary Messenger*, vol. 2, 1836, pp. 318–26.
@repligate. *X*, 23 Oct. 2024, 3.21AM, https://x.com/repligate/status/1848912637091860558
Santos, Janine Patricia. "'We Deserve New Things': (Anti-)Bricolage in Lomé's Makerspaces." *Africa*, vol. 93, 2023, pp. 496–521.
Searle, John R. "Minds, Brains and Programs." *The Behavioral and Brain Sciences*, vol. 3, no. 3, 1980, pp. 417–57.
Shanahan, Murray, and Beth Singler. "Existential Conversations with Large Language Models: Content, Community, and Culture." *arXiv* (2024), https://arxiv.org/abs/2411.13223.
Simondon, Gilbert. *On the Mode of Existence of Technical Objects*. Translated by Cécile Malaspina and John Rogrove, Univocal, 2017.
Singler, Beth. *Religion and Artificial Intelligence: An Introduction*. Routledge, 2024.
@thegoodtimeline. "More info on the S.A.N. (@MycelialOracle) AI project." *X*, 22 Oct. 2024, https://x.com/thegoodtimeline/status/1848544484218572849.
Turkle, Sherry. *Life on the Screen: Identity in the Age of the Internet*. Simon and Schuster, 1995.
Whitehead, Alfred North. *Science and the Modern World*. Cambridge UP, 1953.
Wiggins, Bradley Earl. "The Backrooms and Liminal Spaces: Explorations of a Digital Urban Legend." *New Media & Society*, 2024, pp. 1–29, https://doi.org/10.1177/14614448241238395.
Wittgenstein, Ludwig. *Philosophical Investigations*. Translated by G. E. M. Anscombe. 3rd ed., with index. Basil Blackwell, 1986.

10 Coda

Molly Crozier

The starting point for this volume was a conference which we held at King's College London in January 2023. Having discovered a shared interest in prosthesis as a point of interaction between the human and the non-human, we brought together a group of scholars to address the subject of automation from a range of perspectives. Some of those who attended appear in these pages, others have published that work elsewhere, and in other cases it is still ongoing. In the years that have passed since then, it seems that thinking around automation has drastically altered.

The month of the original conference, January 2023, saw OpenAI's ChatGPT reach 100 million users just two months after its launch, making it the fastest-growing consumer application ever (Milmo). Since then, the discourse about automation and artificial intelligence has been dominated by anxiety. In that same year, Hollywood writers went on strike for nearly five months, with the use of AI in the writing process being one of the key issues at stake, including the use of writers' material for training AI. Perhaps due to the simultaneous actors' strike which saw well-known faces appear on picket lines, this led to heightened awareness of the potential issues AI posed for those in creative careers. It is vital to note that these are risks of which workers in blue-collar jobs whose roles could increasingly be performed mechanically had long been aware (a number of such examples can be found in the introduction to this volume). In our own working lives in universities, AI, the possibility of students using it to cheat and the potential benefits for researchers have become topics discussed with increasing frequency by staff, both among ourselves and in formal training settings, and institutions have rapidly introduced policies addressing the use of AI by students and staff (Dabis and Csáki 2024). In the time that we have been working on this volume, the topic has gone from being a subject of academic interest – a conversation that we had primarily in research contexts – to a conversation that is happening to and around us on a regular basis in our professional, social and online lives. AI shifted from the territory of science fiction, calling to mind humanoid robots (often in the

DOI: 10.4324/9781003545910-11

form of beautiful women) to part of the fabric of our lives, taking the form of (yet more) text on a screen. Even as I am writing this, Microsoft Copilot is a ghostly presence in the document, the icon appearing and disappearing on the left of my screen.

And yet, as much as our own anxiety around automation has heightened during the time we have been working on this volume, the contributions within the collection are a testament to the enduring nature of that anxiety. Humans have always worried about technological advancement – which is not to say that there is no cause for alarm: there is. Rather, it is our contention that there are remedies for that anxiety to be found in the literature and culture of the past, and in continued engagement with ongoing cultural production around the subject of automation. Collectively, the chapters of *The Human and the Machine in Literature and Culture* suggest that our contemporary, anxiety-fuelled debate around the future of automation represents not an unprecedented diversion, a single hinge point in the history of humanity, but the continuation of a conversation that has recurred over a period of centuries. This discourse might have intensified as technological progress has accelerated (or at least, feels like it has), and there is a case to be made that it is more urgent now than it has ever been, but there is no clearly identifiable starting point. Moreover, the chapters of this collection demonstrate that looking at humanity's historical concerns about automation and technology helps us understand the concerns that occupy us at the present moment. We keep worrying about the same things – jobs, loss of autonomy – so it seems worth asking ourselves whether these concerns say more about human beings than they do about technology. Perhaps the past might show us what it is that we are *really* worried about, and even help us to address that.

The collection is by no means comprehensive: no single volume could cover this topic in its entirety, and the rapid progress of automation tools makes this a highly generative subject matter. The sheer number of potential avenues for future research could be somewhat overwhelming, even limiting ourselves to looking to literature and culture of the past. Indeed, a significant risk of this type of research is that there is so much hyperbolic discourse around AI and automation and there is such rapid technological progress that it is all but impossible to predict where the path we are presently on might end. What is a reasonable cause for concern about which action needs to be taken, and what is a fear-driven fantasy? Increasingly, it is hard to tell the difference.

Nonetheless, continued research in this area is essential precisely because of the challenges it poses, and we therefore end this volume with a gesture to the future and other possible avenues for research on the topic. This is research that is ongoing; there was already a healthy body of humanities scholarship on technology, automation and the non-human prior to the drastic uptick in public interest over the last two or three years, and that body of scholarship has only been enriched by that heightened engagement.

As editors, we are conscious that this volume does not address certain time periods, notably the early-modern period. Given the rapid increase in the production of printed books during this time as a result of the invention of the printing press, scholarship engaging with cultural representations of technology at this time might offer valuable insight into our present moment. As Genevieve Warwick and Richard Taws note, 'If our own information age is defined by the digital structures of electronic communication, early modern culture was inextricably bound to the medium of print' (Warwick and Taws, 201). Interested readers might look to Jennifer Oliver's work on war machines in Rabelais's *Le Quart Livre*, Nina Engelhardt's comparison of Francis Godwin's *The Man in the Moone; or a Voyage Thither* and Thomas Pynchon's *Gravity's Rainbow*, or Warwick and Taws's special issue of *Art History*, 'Art and Technology in Early Modern Europe' as starting points.

Similarly, the chapters do not engage substantially with film, which has long been an area for engagement with automation, particularly with the figure of the android or cyborg – although there is, of course, a substantial body of scholarship on this subject. Readers might consider Despina Kakoudaki's *Anatomy of a Robot: Literature, Cinema and the Work of Artificial People*, Eric G. Wilson's *The Melancholy Android: On the Psychology of Sacred Machines*, J.P. Telotte's *Replications: A Robotic History of the Science Fiction Film*, Julie Wosk's *My Fair Ladies: Female Robots, Androids and Other Artificial Eves*, Anne Balsamo's *Technologies of the Cyborg Body: Reading Cyborg Women* and Claudia Springer's *Electronic Eros: Bodies and Desire in the Postindustrial Age* as points of departure. Of particular interest is what seems to be a shift from androids and AI as objects of (sexual or romantic) desire, in films like *Blade Runner*, *Her*, and *Ex Machina* or alternatively threatening (male) androids as in *Terminator*, *Alien*, and *I, Robot*, to sentient robots and technology as a new type of monster in horror films like *M3GAN* and *Five Nights at Freddy's*.

A key area for further investigation must be automation outside of European and American contexts. As Bruno Ministro notes in his contribution to this volume, automation can serve to conceal labour, labour which is often performed by badly paid workers in less economically developed countries. Technoliberals might want automation to deliver us to a future free from work, but the whole notion of an autonomous subject free from work is based on a Eurocentric, post-Enlightenment subject who himself enjoyed a surplus because of the invisibilized work of women, people of colour and enslaved people. Although the replacement of human labour with technology might seem to offer an alternative to racialized and gendered structures, as Neda Atanasoski and Kalindi Vora note, it may actually reinforce them:

> Contrary to the seeming abandonment of the politics of difference in the so-called postrace and postlabor future projected by technoliberal discourses of machine-induced human obsolescence, we thus draw attention to the composition of the human as an abstract category whose expansive capacities continually reaffirm the racial order of things that undergirds Euro-American modernity.
>
> (Atanasoski and Vora 2019, 5)

With this in mind, it is essential that scholarship looking to literature and culture for perspectives on automation considers that which comes from those who are most directly affected. This is all the more important when we note the substantial environmental impact of generative AI, as certain nations (and communities within those nations) are more directly impacted by the effects of the climate crisis. Technological advancement and mechanization form part of the methodology of colonialism, meaning that technological change takes on a new dimension in this context. A lacuna in this volume of which we are particularly conscious is the absence of Afrofuturism among the objects of analysis.

It is our hope that holding the subject matter at something of a remove, whether that be due to temporal distance or the intellectual distance offered by couching the discussion in cultural representation, allows for a more nuanced engagement, moving beyond polarized responses. This, we believe, is the reason that future research in this area is essential, and ought to be the methodology underpinning any such research: neither to deify nor to villainize automation, but to regard it in its complexity, in the way that literature and culture enable us to do.

March 2025

References

Atanasoski, Neda, and Kalindi Vora. *Surrogate Humanity: Race, Robots, and the Politics of Technological Futures*. Duke UP, 2019.

Dabis, Attila, and Csaba Csáki. "AI and Ethics: Investigating the First Policy Responses of Higher Education Institutions to the Challenge of Generative AI." *Humanities and Social Sciences Communications*, vol. 11, no. 1, 6 Aug. 2024, https://doi.org/10.1057/s41599-024-03526-z

Engelhardt, Nina. "'Real Flight and Dreams of Flight Go Together': High Technology and Imaginary Heights in Early Modern and Postmodern Science Fiction." *Space and Culture*, vol. 23, no. 4, 25 Dec. 2018, pp. 382–93.

Milmo, Dan. "ChatGPT Reaches 100 Million Users Two Months after Launch." *The Guardian*, 2 Feb. 2023, https://www.theguardian.com/technology/2023

/feb/02/chatgpt-100-million-users-open-ai-fastest-growing-app/. Accessed 19 Mar. 2025.

Oliver, Jennifer. “Rabelais’s Engins: War Machines, Analogy, and the Anxiety of Invention in the Quart Livre.” *Early Modern French Studies*, vol. 38, no. 2, 2 Jul. 2016, pp. 97–110.

Warwick, Genevieve, and Richard Taws. “After Prometheus: Art and Technology in Early Modern Europe.” *Art History*, vol. 39, no. 2, 23 Mar. 2016, pp. 198–209.

Index

Note: 'n' denotes footnote numbers

Kolibri 360 GmbH
Ferdinand-Jühlke-Straße 7
99095 Erfurt, Deutschland
produktsicherheit@kolibri360.de